CAMBRIDGE TRACTS IN MATHEMATICS

General Editors

J. BERTOIN, B. BOLLOBÁS, W. FULTON, B. KRA, I. MOERDIJK,
C. PRAEGER, P. SARNAK, B. SIMON, B. TOTARO

CAMBRIDGE TRACTS IN MATHEMATICS

GENERAL EDITORS

J. BERTOIN, B. BOLLOBÁS, W. FULTON, B. KRA, I. MOERDIJK,
C. PRAEGER, P. SARNAK, B. SIMON, B. TOTARO

A complete list of books in the series can be found at www.cambridge.org/mathematics.
Recent titles include the following:

Transcendence and Linear Relations of 1-Periods

ANNETTE HUBER
Albert-Ludwigs-Universität Freiburg, Germany

GISBERT WÜSTHOLZ
ETH Zürich

CAMBRIDGE
UNIVERSITY PRESS

CAMBRIDGE
UNIVERSITY PRESS

University Printing House, Cambridge CB2 8BS, United Kingdom

One Liberty Plaza, 20th Floor, New York, NY 10006, USA

477 Williamstown Road, Port Melbourne, VIC 3207, Australia

314–321, 3rd Floor, Plot 3, Splendor Forum, Jasola District Centre,
New Delhi – 110025, India

103 Penang Road, #05–06/07, Visioncrest Commercial, Singapore 238467

Cambridge University Press is part of the University of Cambridge.

It furthers the University's mission by disseminating knowledge in the pursuit of
education, learning, and research at the highest international levels of excellence.

www.cambridge.org
Information on this title: www.cambridge.org/9781316519936
DOI: 10.1017/9781009019729

First published 2022

A catalogue record for this publication is available from the British Library.

Library of Congress Cataloging-in-Publication Data
Names: Huber, Annette, author. | Wüstholz, Gisbert, author.
Title: Transcendence and linear relations of 1-periods / Annette Huber, Gisbert Wüstholz.
Description: Cambridge ; New York, NY : Cambridge University Press, 2022. |
Series: Cambridge tracts in mathematics ; 227 |
Includes bibliographical references and index.
Identifiers: LCCN 2022002180 | ISBN 9781316519936 (hardback)
Subjects: LCSH: Transcendental numbers. | Motives (Mathematics) | Algebraic fields. |
BISAC: MATHEMATICS / Number Theory
Classification: LCC QA247.5 .H83 2022 | DDC 512.7/3–dc23/eng20220318
LC record available at https://lccn.loc.gov/2022002180

ISBN 978-1-316-51993-6 Hardback

In memoriam Alan Baker

Contents

Prologue

The study of transcendence properties of periods has a long history. It began in 1882 with the famous theorem of Lindemann on the transcendence of π, which showed that squaring the circle is not possible. This settled a problem more than 2000 years old from the time of the Greeks. At the same time, he showed that α and e^α cannot both be algebraic unless $\alpha = 0$. In particular, $\log \alpha$ is transcendental for algebraic $\alpha \neq 0, 1$. Lindemann actually proved more: his method gives us that if $\alpha_1, \ldots, \alpha_N$ are pairwise distinct algebraic numbers then $e^{\alpha_1}, \ldots, e^{\alpha_N}$ are linearly independent over $\overline{\mathbb{Q}}$. This was carried out in full detail, and with the approval of Lindemann, in 1885 by Weierstraß in [Wei85, p. 1067, footnote 2].

In his famous address at the ICM 1900, Hilbert went further. In the seventh of his 23 problems, he asked when α, β and $\gamma = \alpha^\beta$ can all three be algebraic numbers? There are some obvious cases where this is true, namely when $\alpha = 0$, $\alpha = 1$ or β is rational. But Hilbert went further and asked whether these were the only cases. He considered this problem as more difficult to prove than the Riemann Hypothesis.

To much surprise, Gelfond [Gel34] and Schneider [Sch34b], applying different methods, independently succeeded in 1934 in answering Hilbert's Problem. An almost equivalent formulation is that for $\alpha \neq 0$ the three numbers are algebraic if $\log(\alpha)$ and $\log(\alpha^\beta)$ are linearly dependent over $\mathbb{Q}$.

Linear Forms of Logarithms

The work of Gelfond and Schneider initiated extensive work on so-called linear forms in logarithms. It was known that lower bounds for such linear forms would give solutions to several outstanding problems. One of them is the famous class number 1 problem; another is finding effectively the integral solutions of classes of diophantine equations. The main open problem in this context was to deal with linear forms in three logarithms of algebraic numbers with algebraic coefficients.

The methods that had been developed so far could not handle more than two logarithms and it was considered a very difficult problem to make progress on. This hurdle was overcome in the case of classical logarithms in 1966 by Baker. In his famous paper he was able to solve this problem in full generality not only for three but even for any finite number of logarithms. In particular, he showed that if $\alpha_1, \ldots, \alpha_n$ are non-zero algebraic numbers, then a linear form with algebraic coefficients in $\log \alpha_1, \ldots, \log \alpha_n$ vanishes if and only if the logarithms are linearly dependent over $\mathbb{Q}$. This result is exactly in the spirit of Gelfond and Schneider's solutions of the Hilbert Problem.

Elliptic and Abelian Integrals

The number $2\pi i$ is a period of the integral $\int dx/x$ of the rational differential form dx/x taken over a closed path in $\mathbb{C}$. Other period numbers appear in the theory of the Weierstraß elliptic function as elliptic integrals of the first kind, and, consequently, it was no surprise that Siegel [Sie32] took up the topic. He proved that not all periods of the Weierstraß elliptic functions with algebraic invariants g_2 and g_3 are algebraic; in particular it follows that in the complex multiplication case all non-zero periods are transcendental.

Schneider further developed the theory to deal also with elliptic integrals of the first and second kind and complete or incomplete periods. In a series of papers based on his solution of Hilbert's seventh problem [Sch34b, Sch34a], he went as far as the technical tools of the time allowed. He proved in [Sch37], for example, that if u is chosen such that the Weierstraß $\wp$-function takes an algebraic value, then 1, u and $\zeta(u)$ are linearly independent over $\overline{\mathbb{Q}}$. In particular, if ω is in the period lattice then 1, ω and $\eta(\omega)$ are linearly independent over $\overline{\mathbb{Q}}$. This was the first paper in which he proved a result about the Weiserstrass $\wp$-function and the exponential function. In particular he showed that π/ω is transcendental.

In the subsequent years Schneider extended his work, studying the transcendence properties of abelian functions and integrals and obtaining the first, albeit partial, results. The most striking example was the transcendence of the values of the B-function at rational arguments; see also [Sie49]. In his book [Sch57], Schneider asked, as open problems, for proofs of similar results for elliptic integrals of the third kind and for abelian integrals. As he noted, it was clear that the methods were exhausted and new methods would be necessary to solve the problems.

In a series of papers following his work on logarithms, Baker was also able to extend in a very limited way Schneider's results about transcendence of values of elliptic functions in the case of the Weierstraß $\wp$- and ζ-functions which also contain transcendence results on elliptic integrals of the second kind; see, for example, [Bak69]. His results were extended to linear independence results on periods of integrals of the second kind by himself, Coates, Masser, Laurent and Bertrand. However, they were also strongly limited by the lack of a general tool for handling the case of an arbitrary number of elliptic or abelian logarithms. The problem was that Baker's approach did not work in general in the case of abelian varieties or more generally in the case of commutative algebraic groups.

This changed only when the Analytic Subgroup Theorem [Wüs89] became available in 1982. It allowed one to deal with 1-periods in general and linear relations between them. The case of *complete* periods in the general case where ω is an algebraic 1-form on a curve of arbitrary genus and γ a closed path on the corresponding Riemann surface was settled in 1986 by the second author in [Wüs87]: if a period is non-zero, it is transcendental. Both cases can arise. A simple example is a hyperelliptic curve whose Jacobian is isogenous to a product of two elliptic curves. Then 8 of the 16 standard periods are 0. The others are transcendental.

The Analytic Subgroup Theorem opened a fruitful interplay between transcendence theory and algebraic groups through the exponential map for Lie groups. More recently it turned out that the right frame is the theory of 1-motives introduced by Deligne in 1974; see [Del74]. It added to the algebraic groups extra data in the shape of a homomorphism of a free abelian group into the group, which lets one deal with so-called *incomplete* periods in an elegant and natural way. We develop this point of view in full detail in the present monograph.

Grothendieck's Period Conjecture

The transcendence properties of periods make it natural to ask questions about linear and algebraic relations between them. A conceptual interpretation of possible relations is provided by what came to be known as the *Period Conjecture*. Periods are given a cohomological interpretation and all relations between them should be induced by relations between motives.

This string of ideas was started by Grothendieck in [Gro66, p. 101]. He discussed the comparison of the de Rham cohomology of a smooth variety X over a number field K with its singular cohomology. The entries of the comparison matrix comparing $H^1_{dR}(X)$ and $H^1_{sing}(X^{an}, \mathbb{Q})$ of a complete non-singular curve X are classical periods of the first and second kind. Grothendieck asks for instance if Schneider's Theorem generalises in some way to these periods.[1] Subsequently, he came to a conceptual conjecture and went further by predicting the transcendence degree of the field of periods of $H^n(X)$ for a smooth projective variety X (or more generally of a pure motive M) as the dimension of the motivic Galois group or alternatively the dimension of the Mumford–Tate group of the Hodge structure on M.

However, he did not publish the conjecture himself. We refer to the first-hand account of André in [And19] on the history of the conjecture. A complete formulation and discussion was finally given by André [And04, Chapter 23]. This includes a straightforward extension of the conjecture to the case of mixed motives. The formulation of the conjecture for 1-motives is discussed by Bertolin in [Ber02] and more recently [Ber20]. The only known result in this direction is a theorem of Chudnovsky, who showed in [Chu80] that for any elliptic curve defined over $\overline{\mathbb{Q}}$ at least two of the numbers ω_1, ω_2, $\eta(\omega_1)$ and $\eta(\omega_2)$ are algebraically independent provided that ω_1 and ω_2 generate the period lattice over $\mathbb{Q}$. In the case of complex multiplication this implies that ω and $\eta(\omega)$ are algebraically independent, which confirms the prediction in the CM-case. In general Grothendieck's Conjecture is out of reach. We shall present an overview of further developments below.

The Period Conjecture as Formulated by Kontsevich and Zagier

In a series of papers, Kontsevich and Zagier [KZ01, Kon99] also promoted the study of periods of non-smooth, non-projective varieties, or more generally of mixed motives. In [Kon99] Kontsevich formulated an alternative version of the Period Conjecture. As André pointed out, it has a very different flavour based on calculus rather than algebraic geometry. Relations between periods are induced by the transformation rule and Stokes's Theorem. This approach puts relative cohomology front and centre. Kontsevich views periods as the

---

[1] In footnote 10 of [Gro66] Grothendieck recalls the belief that the periods ω_1, ω_2 of a non-CM elliptic curve should be algebraically independent. 'This conjecture extends in an obvious way to the set of periods $(\omega_1, \omega_2, \eta_1, \eta_2)$ and can be rephrased also for curves of any genus, or rather for abelian varieties of dimension g, involving $4g$ periods.'

numbers in the image of the period pairing given by period integrals for relative cohomology

$$H_{\mathrm{dR}}^*(X, Y) \times H_*^{\mathrm{sing}}(X, Y; \mathbb{Q}) \to \mathbb{C}$$

for algebraic varieties X over $\overline{\mathbb{Q}}$ and subvarieties $Y \subset X$. By *Kontsevich's Period Conjecture* all $\overline{\mathbb{Q}}$-linear relations between such periods should be induced by bilinearity and functoriality of mixed motives. More explicitly, he introduces an algebra of formal periods $\widetilde{\mathcal{P}}$ (also called motivic periods by some authors) with explicit generators and relations. His conjecture predicts that the evaluation map $\widetilde{\mathcal{P}} \to \mathbb{C}$ (sending a formal period to the actual value of the integral) is injective. In the present monograph, we give an answer in the case of periods in degree 1, or, equivalently, periods of curves.

Grothendieck's Period Conjecture on algebraic relations between periods is essentially equivalent to Kontsevich's Period Conjecture on linear relations between periods; see, for example, [Ayo14a], [And17], [HMS17, Section 13.2.1] or [Hub20, Section 5.3] for the precise relation. It rests on a key insight of Nori, who realised that $\mathrm{Spec}(\widetilde{\mathcal{P}})$ of the algebra $\widetilde{\mathcal{P}}$ is a torsor under a motivic Galois group, in fact the torsor of tensor isomorphisms between the de Rham realisation and the singular realisation of the category of mixed Nori motives.

Dimensions of Period Spaces

For any given variety, the space of periods is finite dimensional. This makes it natural to ask about their dimension. A qualitative prediction already follows from the Period Conjecture. The precedent for the kind of formula that we have in mind is Baker's Theorem [Bak66] on logarithms. Such a formula can be made explicit: for $\beta_1, \ldots, \beta_n \in \overline{\mathbb{Q}}^*$ let $\langle \beta_1, \ldots, \beta_n \rangle$ be the multiplicative subgroup of $\overline{\mathbb{Q}}^*$ generated by these numbers. Then the dimension of the vector space generated by the principal determinations of $\log \beta_1, \ldots, \log \beta_n$ over $\overline{\mathbb{Q}}$ (but modulo multiples of π) is equal to the rank of the group generated by $\beta_1, \ldots, \beta_n$.

In addition to Baker's Theorem, a number of cases have been considered in the past: for example, the case of elliptic logarithms in [BW07, Section 6.2] or the extension of an elliptic curve by a torus of dimension n in [Wüs84b]. An interesting new case came up recently in connection with curvature lines and geodesics for billiards on a triaxial ellipsoid; see [Wüs21]. This leads to a period space generated by 1, $2\pi i$ and the periods $\omega_1, \omega_2, \eta(\omega_1), \eta(\omega_2), \lambda(u, \omega_1), \lambda(u, \omega_2)$ of the first, second and third kind. Its dimension over $\overline{\mathbb{Q}}$ is 8, 6 or 4 depending on the endomorphisms of the elliptic curve involved and on the nature of the differential of the third kind. This case serves as a model for a completely general result.

In the case of 1-motives this is the question about the dimension of the vector space generated over $\overline{\mathbb{Q}}$ by their periods. It turns out that to state and prove such a general formula for the dimension of the period space of a 1-motive is difficult. The difficulties arise from periods of the third kind and the formulas we shall give are quite involved.

Outlook

As we have already stated, the Period Conjecture itself seems currently far out of reach. Even the special case of values of the Riemann ζ-functions is widely open. The Period Conjecture implies that the $\zeta(2n + 1)$ for $n \in \mathbb{N}$ are algebraically independent. On the one hand we have the theory of motives and on the other hand transcendence theory. The interaction between both is a wonderful topic. It has been exploited in order to deduce upper bounds for the spaces of periods. However, on the transcendence side, only comparatively weak lower bounds are available. Only for the case of 1-motives over $\overline{\mathbb{Q}}$ do we have a complete description of the transcendental aspects. This is what our book explains.

It is appropriate to give a short overview of other types of motives which were studied with respect to transcendence. This concerns motives over $\mathbb{Q}$, motives over function fields, both over $\mathbb{Q}$ as well as over finite fields. The more structure the base field has the more complete the transcendence situation becomes. In the following we go through some cases of motives for which transcendence has been studied and we also mention some problems about effectivity.

Mixed Tate Motives

The Riemann ζ-function has been, since the work of Euler, one of the central objects in number theory. Euler showed that its value at positive integers $2n$ is, up to a constant, of the form $(2\pi)^{2n}$. This implies that its values are transcendental over the rationals. This is Lindemann's Theorem. The only other known fact about irrationality or transcendence of integral values of the ζ-function is Ápery's discovery of irrationality of $\zeta(3)$; see [Apé79]. Only since about 2000 has there been more intensive study of these values, starting with Rivoal, Zudilin and others; see, for example, [Zud01], [BR01]. They considered the space over $\mathbb{Q}$ generated by odd ζ values up to a fixed integer n for small n and first showed that its dimension is at least 1. More recently they have been able to prove that the dimension tends to infinity with n at a rate of at least order $\log n$.

From the other side, upper bounds on spaces generated by ζ and multi-zeta values (the periods of mixed Tate motives over $\mathbb{Z}$) were provided by Deligne–Goncharov in [DG05]. The reason why these results could be established is that the motivic picture is completely understood; see also Brown's work [Bro12] on the structure of the motivic Galois group in this case.

Motives over Function Fields over $\mathbb{Q}$

Ayoub reformulated Kontsevich's Period Conjecture with fewer generators. Only polydisks are needed as domains of integration. Based on this description, he was able to formulate and prove a function field version of the conjecture in [Ayo15].

For a closed polydisk $\overline{\mathbb{D}}^n$ he considered the subspace

$$\mathcal{O}^{\dagger}_{\mathrm{alg}}(\overline{\mathbb{D}}^n) \subset \mathcal{O}(\overline{\mathbb{D}}^n)[[T]][T^{-1}]$$

of Laurent series $F = \sum_{i > -\infty} f_i(z_1, \ldots, z_n) T^i$ with coefficients in $\mathcal{O}(\overline{\mathbb{D}}^n)$ which are algebraic over $\mathbb{C}(T, z_1, \ldots, z_n)$. The dimension n is allowed to vary and $\mathcal{O}^{\dagger}_{\mathrm{alg}}(\overline{\mathbb{D}}^\infty) = \bigcup_{n \in \mathbb{N}} \mathcal{O}^{\dagger}_{\mathrm{alg}}(\overline{\mathbb{D}}^n)$. In analogy to Kontsevich's space of formal periods $\widetilde{\mathcal{P}}$, he defined $\mathcal{P}^{\dagger}$ as a quotient of $\mathcal{O}^{\dagger}_{\mathrm{alg}}(\overline{\mathbb{D}}^\infty)$ by certain relations and showed that there is an evaluation map $\mathcal{P}^{\dagger} \to \mathbb{C}((T))$. The main result (and geometric analogue of the Period Conjecture) is the injectivity of this evaluation map. There is also independent work of Nori (unpublished) in the same direction.

Motives over Function Fields over Finite Fields

All that is known about transcendence over $\mathbb{Q}$ is also known for function fields over finite fields: indeed often more is known. Let p be a prime, $q = p^n$ and $\mathbb{F}_q[x]$ the ring of polynomials in one variable over the finite field $\mathbb{F}_q$. In 1935 Carlitz introduced the so-called Carlitz ψ-functions in [Car35], defined as

$$\psi(t) = \sum_0^\infty \frac{(-1)^k}{F_k} t^{q^k},$$

where $F_k = [k][k-1]^q \cdots [1]^{q^{k-1}}$ and $[k] = x^{q^k} - x$. On the basis of these functions, Wade started studying transcendence theory over function fields over finite fields [Wad46]. Their minimal algebraic closure is complete with respect to the standard valuation so that the function $\psi(x)$ exists and is a replacement of the exponential function. Its inverse exists as a multi-valued function and is the analogue of the logarithm. Wade proved among other things the analogue of the theorem of Gelfond and Schneider.

In 1983 Jing Yu took up the topic and proved the analogue of Lindemann's Theorem in the realm of Drinfeld modules and started a very interesting transcendence theory for Drinfeld modules and t-motives. As a highlight of a sequence of papers including periods and quasi-periods of Drinfeld modules as well as special zeta values in characteristic p, he obtained an analogue of the Analytic Subgroup Theorem for Drinfeld modules and, more generally, Anderson's t-motives. Once the theory and the techniques had been established, a whole spectrum of applications followed, including linear independence of zeta values, by Yu, Chieh-Yu Chang, Papanikolas and Thakur; see [Yu97], [CY07], [CPTY10]. They were even able to determine the transcendence degree of fields generated by logarithms or zeta values in this setting. The survey paper of Chang [Cha17] gives a very nice and substantial report about the newest achievements in this theory.

Hypergeometric Period Relations, Periods of Higher Weight

It is well known that values of hypergeometric functions can be expressed as a quotient of two abelian integrals, in general of the second kind. This leads to a period relation between the two periods of the second kind with the hypergeometric function as coefficient. Algebraic values of the hypergeometric function provide linear relations between the two periods with algebraic coefficients. This cannot be true in general and leads to special points on certain Shimura varieties as explained very carefully in Chapter 5 of Tretkoff's beautiful monograph [Tre17]. In particular, new transcendence results are given for the Appell–Lauricella (hypergeometric) functions in $n \geq 2$ variables. They exceed the known results on the values of the classical hypergeometric function in one variable.

Hodge Level 1

An obvious problem is to extend the transcendence results to periods of higher weight. In general this seems to be a hopeless undertaking. However, there are cases when periods of higher weight can be related to 1-periods. Tretkoff gives some nice examples dealing with periods of Fermat hypersurfaces. This is the case for certain algebraic $K3$-surfaces and smooth complete intersections over $\mathbb{C}$ of Hodge level 1 as explained in [Del72]; see also [Wüs87].

We are not going to expand our monograph in this direction but mention some interesting research dealing with this kind of problem. The starting point was given in [Wüs87] and then taken up by Tretkoff.

In Chapter 6 of her monograph [Tre17], P. Tretkoff discusses among other things algebraic K3-surfaces X defined over a number field. She considers a holomorphic 2-form ω on X and shows that if the vector space generated by the periods $\int_\gamma \omega$ for $\gamma \in H_2(X, \mathbb{Z})$ has dimension 1, then X has complex multiplication, i.e. its Mumford–Tate group is abelian. This implies that if X has complex multiplication these periods are all transcendental unless $\gamma = 0$.

In Chapter 7 Tretkoff deals with arbitrary smooth projective varieties X defined over $\overline{\mathbb{Q}}$. One of the questions she raises is whether the Hodge filtration of $H^k(X, \mathbb{C})$ for $0 \leq k \leq \dim X$ has complex multiplication if it is defined over $\overline{\mathbb{Q}}$. She gives some nice examples dealing with periods on Fermat hypersurfaces.

All this is restricted to holomorphic differential forms and complete periods. In our monograph we deal with meromorphic differential forms and incomplete periods. It would be interesting to try to get more general cases involving incomplete periods.

Effectivity, Lower Bounds

The main applications of Baker's work on linear forms in logarithms were the lower bounds he derived. He showed that if Λ is a linear form in logarithms of algebraic numbers with algebraic coefficients, a lower bound for the absolute value of Λ can be obained. For a detailed account on this, see [BW07]. A similar theory exists also for the p-adic analogue, started by Coates and finally brought to the level of the archimedean case by Kunrui Yu. Similar results were also obtained for elliptic and abelian logarithms. One might ask whether this can be extended to 1-motives in a modified way. First steps in this direction were the work of Masser and Wüstholz [MW93] on isogeny estimates. It says that, given an isogeny between two abelian varieties, there exists an isogeny with degree bounded by the original data: height of the source isogeny, degree of the number field, dimension of the abelian variety. The result is completely effective and a crucial ingredient for the proof of the famous Tate Conjecture. It has also been used for the proof of the André–Oort Conjecture for the coarse moduli space of prinicipally polarised abelian varieties by Tsimerman [Tsi18]. It would be interesting to formulate an isogeny estimate type statement for 1-motives and to find applications in diophantine geometry.

Acknowledgements

The two authors gratefully acknowledge the hospitality of the Freiburg Institute for Advanced Study, in particular the Research Focus Cohomology in Algebraic Geometry and Representation Theory, in the academic year 2017/18. Most of the book was written during this stay.

We thank Fritz Hörmann, Florian Ivorra and Simon Pepin-Lehalleur for discussions on 1-motives and the generalised Jacobian. We also thank Cristiana Bertolin for comments on an earlier version and Jürgen Wolfart for help with hypergeometric series. Alin Bostan explained the corrected proof of the explicit formula for values of the hypergeometric series in Section 19.1.1 to us. The argument for an alternative proof of Theorem 13.9 was pointed out to us by David Masser. We explain a variant at the end of the chapter.

We are grateful to Yves André for his historical comments, reading some part of the manuscript in detail and graciously sharing his insights on the Period Conjecture.

Finally, the first author would like thank all participants of her 2020 lecture on the Analytic Subgroup Theorem, in particular Fabian Glöckle, for their lively interest, comments and corrections.

1

Introduction

This introduction aims to present the principal actors of the book and to explain the main results of our monograph. We begin with the question about transcendence of periods of integrals of 1-forms over closed or non-closed paths. Historically integrals over non-closed paths were not considered as periods. The change came from looking at them in the relative cohomology. This leads us to distinguish between complete and incomplete periods.

The order of topics presented here does not follow the order in the main text but is, we hope, designed to help those readers without a background in transcendence.

1.1 Transcendence

The vector space $\mathcal{P}^1$ over $\overline{\mathbb{Q}}$ of one-dimensional periods, complete or incomplete, has a number of different descriptions. In the most elementary situation its elements are given by the period integrals

$$\alpha = \int_\sigma \omega,$$

where

- X is a smooth projective curve over $\overline{\mathbb{Q}}$;
- ω is a rational differential form on X;
- $\sigma = \sum_{i=1}^n a_i \gamma_i$ is a chain in the Riemann surface X^{an} defined by X which avoids the singularities of ω and has boundary divisor $\partial\sigma$ in $X(\overline{\mathbb{Q}})$; in particular $\gamma_i \colon [0,1] \to X^{\mathrm{an}}$ is a path and $a_i \in \mathbb{Z}$.

This set includes many interesting numbers like $2\pi i$, $\log \alpha$ for algebraic α and the periods of elliptic curves over $\overline{\mathbb{Q}}$. We study their transcendence properties.

The case of *complete* periods in the general case, i.e. X and ω arbitrary, γ closed, was settled in 1986 by the second author in [Wüs87]: if a period is non-zero, it is transcendental. Both cases can arise. A simple example is a hyperelliptic curve whose Jacobian is isogenous to a product of two elliptic curves. Then 8 of the 16 standard periods are 0. The others are transcendental.

When X is an elliptic curve we refer the reader to [BW07, Section 6.2] for the case of *incomplete* periods. The general case has been described as an open problem in [Wüs84a]. Often the values are transcendental, e.g. $\int_1^2 dz/z = \log 2$, but certainly not always, e.g. $\int_0^2 dz = 2$. Again, it is not difficult to write down a list of simple cases in which the period is a non-zero algebraic number. However, it was not at all clear whether the list was complete and what the structure behind the examples was; see [Wüs12]. The answer that we give now is surprisingly simple:

Theorem 1.1 (Theorem 13.9). *Let* $\alpha = \int_\sigma \omega$ *be a one-dimensional period on* X. *Then* α *is algebraic if and only if*

$$\omega = df + \omega',$$

where $f \in \overline{\mathbb{Q}}(X)^*$ *and* $\int_\sigma \omega' = 0$ *with* ω' *a form with no extra poles.*

The condition is clearly sufficient because the integral evaluates to

$$\sum_i a_i(f(\gamma_i(1)) - f(\gamma_i(0))) \in \overline{\mathbb{Q}}$$

in this case.

Theorem 13.9 gives a complete answer to two of the seven problems listed in Schneider's book [Sch57, p. 138], [1] [2] open for more than 60 years. We even include periods of abelian integrals of the third kind.

1.2 Relations Between Periods

Questions on transcendence can be viewed as a very special case of the question on $\overline{\mathbb{Q}}$-linear relations between 1-periods: a complex number is transcendental if it is $\overline{\mathbb{Q}}$-linearly independent of 1. The most general problem of this kind is to determine the dimension of the period space generated over $\overline{\mathbb{Q}}$ by the periods of all rational 1-forms of an algebraic variety. It is easy to give an upper bound for this dimension in terms of cohomological data. The problem

---

[1] Problem 3. Es ist zu versuchen, Transzendenzresultate über elliptische Integrale dritter Gattung zu beweisen.

[2] Problem 4. Die Transzendenzsätze über elliptische Integrale erster und zweiter Gattung sind in weitestmöglichem Umfang auf analoge Sätze über abelsche Integrale zu verallgemeinern.

is then to decide whether the upper bound is the correct number or whether there are linear relations between periods.

This fundamental question will be one of the central topics in this monograph. We establish a complete description of the linear relations between (not necessarily complete) periods for all rational differential forms of degree 1. It is crucial to use here the more conceptual descriptions of $\mathcal{P}^1$ either as periods in cohomological degree 1 or as cohomological periods of curves, or even better periods of 1-motives.

The following theorem gives a first answer. It establishes Kontsevich's version of the Period Conjecture for $\mathcal{P}^1$ and furnishes a qualitative description of the period relations.

Theorem 1.2 (Kontsevich's Period Conjecture for $\mathcal{P}^1$, Theorem 13.3). *All $\overline{\mathbb{Q}}$-linear relations between elements of $\mathcal{P}^1$ are induced by bilinearity and functoriality of pairs (C, D) where C is a smooth affine curve over $\overline{\mathbb{Q}}$ and $D \subset C$ a finite set of points over $\overline{\mathbb{Q}}$.*

The conjecture has an alternative formulation in terms of motives. In fact, we deduce Theorem 1.2 from the motivic version below, together with the result of Ayoub and Barbieri-Viale in [ABV15] which says that the subcategory of $\mathcal{MM}_{\text{Nori}}^{\text{eff}}$ generated by $H^*(C, D)$ with C of dimension at most 1 agrees with Deligne's much older category of 1-motives; see [Del74].

Every 1-motive M has a singular realisation $V_{\text{sing}}(M)$ and a de Rham realisation $V_{\text{dR}}(M)$. They are linked via a period isomorphism

$$V_{\text{sing}}(M) \otimes_{\mathbb{Q}} \mathbb{C} \cong V_{\text{dR}}(M) \otimes_{\overline{\mathbb{Q}}} \mathbb{C}.$$

There is a well-known relation between curves and 1-motives provided by the theory of generalised Jacobians. From this fact we see that the set $\mathcal{P}^1$ has another alternative description as the union of the images of the period pairings

$$V_{\text{sing}}(M) \times V_{\text{dR}}^{\vee}(M) \to \mathbb{C}$$

for all 1-motives M over $\overline{\mathbb{Q}}$.

Theorem 1.3 (Period Conjecture for 1-motives, Theorem 9.10). *All $\overline{\mathbb{Q}}$-linear relations between elements of $\mathcal{P}^1$ are induced by bilinearity and functoriality for morphisms of iso-1-motives over $\overline{\mathbb{Q}}$.*

This theorem does not say anything about the actual dimension of the period space. We need a quantitative answer. In other words the space of relations has to be determined. It turns out that finding the answer is rather difficult in some cases.

1.3 Dimensions of Period Spaces

The above qualitative theorems can be refined into an explicit computation of the dimension $\delta(M)$ of the $\overline{\mathbb{Q}}$-vector space generated by the periods of a given 1-motive M. The result depends on the subtle and very unexpected interplay between the constituents of M.

Not only for the proofs, but also for the very formulation of the dimension formulas, we rely on the theory of 1-motives introduced by Deligne; see [Del74]. They form an abelian category that captures all cohomological properties of algebraic varieties in degree 1, including all one-dimensional periods.

We review the basics: a 1-motive over $\overline{\mathbb{Q}}$ is a complex $M = [L \to G]$, where G is a semi-abelian variety over $\overline{\mathbb{Q}}$ and L is a free abelian group of finite rank. The map is a group homomorphism. As mentioned earlier, every 1-motive has de Rham and singular realisations, and a period isomorphism between them after extension of scalars to $\mathbb{C}$.

If C is a smooth curve over k, $D \subset C$ a finite set of $\overline{\mathbb{Q}}$-points, then there is a 1-motive $M_1(C)$ such that $H_1^{\mathrm{sing}}(C^{\mathrm{an}}, D; \mathbb{Q})$ agrees with the singular realisation of $M_1(C)$, and $H_{\mathrm{dR}}^1(C, D)^\vee$ agrees with the de Rham realisation of $M_1(C)$. Hence the periods of the pair (C, D) agree with the periods of $M_1(C)$. Explicitly, $M_1(C) = [\mathbb{Z}[D]^0 \to J(C)]$, where $J(C)$ is the generalised Jacobian of C and $\mathbb{Z}[D]^0$ means divisor of degree 0 supported on D.

We denote by $\mathcal{P}(M)$ the image of the period pairing for M and by $\mathcal{P}\langle M \rangle$ the abelian group (or, equivalently, $\overline{\mathbb{Q}}$-vector space) generated by $\mathcal{P}(M) \subset \mathbb{C}$.

We fix a 1-motive $M = [L \to G]$, with G an extension of an abelian variety A by a torus T and L a free abelian group of finite rank. For the definition of its singular realisation $V_{\mathrm{sing}}(M)$ and its de Rham realisation $V_{\mathrm{dR}}^\vee(M)$, we refer the reader to Chapter 8.

The weight filtration on M, explicitly given by

$$[0 \to T] \subset [0 \to G] \subset [L \to G],$$

induces

$$V_{\mathrm{sing}}(T) \hookrightarrow V_{\mathrm{sing}}(G) \hookrightarrow V_{\mathrm{sing}}(M)$$

and dually

$$V_{\mathrm{dR}}^\vee(M) \leftarrow V_{\mathrm{dR}}^\vee([L \to A]) \leftarrow V_{\mathrm{dR}}^\vee([L \to 0]).$$

Together, they introduce a bifiltration

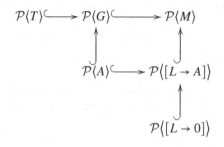

on $\mathcal{P}\langle M\rangle$.

We introduce the following notation and terminology:

$\mathcal{P}_{\mathrm{Ta}}(M) = \mathcal{P}\langle T\rangle$ Tate periods,

$\mathcal{P}_2(M) = \mathcal{P}\langle A\rangle$ 2nd kind wrt closed paths,

$\mathcal{P}_{\mathrm{alg}}(M) = \mathcal{P}\langle[L \to 0]\rangle$ algebraic periods,

$\mathcal{P}_3(M) = \mathcal{P}\langle G\rangle/(\mathcal{P}_{\mathrm{Ta}}(M) + \mathcal{P}_2(M))$ 3rd kind wrt closed paths,

$\mathcal{P}_{\mathrm{inc2}}(M) = \mathcal{P}\langle[L \to A]\rangle/(\mathcal{P}_2(M) + \mathcal{P}_{\mathrm{alg}}(M))$ 2nd kind wrt non-cl. paths,

$\mathcal{P}_{\mathrm{inc3}}(M) = \mathcal{P}\langle M\rangle/(\mathcal{P}_3(M) + \mathcal{P}_{\mathrm{inc2}}(M))$ 3rd kind wrt non-cl. paths,

where wrt and non-cl. are abbreviations for 'with respect to' and 'non-closed'. After choosing bases, we can organise the periods into a period matrix of the form

$$\begin{pmatrix} \mathcal{P}_{\mathrm{Ta}}(M) & \mathcal{P}_3(M) & \mathcal{P}_{\mathrm{inc3}}(M) \\ 0 & \mathcal{P}_2(M) & \mathcal{P}_{\mathrm{inc2}}(M) \\ 0 & 0 & \mathcal{P}_{\mathrm{alg}}(M) \end{pmatrix}.$$

The contribution of $\mathcal{P}_{\mathrm{Ta}}(M)$ (multiples of $2\pi i$) and $\mathcal{P}_{\mathrm{alg}}(M)$ (algebraic numbers) is readily understood. Note that the off-diagonal entries are only well defined up to periods on the diagonal. This can also be seen in the case of Baker periods, which are contained in $\mathcal{P}_{\mathrm{inc3}}(M)$ for special M. The value of $\log \alpha$ depends on the chosen path and is only well defined up to multiples of $2\pi i$. The total dimension is obtained by adding up these dimensions. In particular, we have, for example,

$$\mathcal{P}\langle[L \to A]\rangle \cap \mathcal{P}\langle[0 \to G]\rangle = \mathcal{P}\langle[0 \to A]\rangle.$$

The complete result takes a rather complicated form. In order to state it we write $\delta(M) = \dim \mathcal{P}\langle M\rangle$ and $\delta_?(M) = \dim \mathcal{P}_?(M)$ for the different entries of the period matrix. If B is a simple abelian variety, $g(B)$ will be its genus and

$e(B)$ the $\mathbb{Q}$-dimension of $\text{End}(B)_{\mathbb{Q}}$. We also need the invariants $\text{rk}_B(L, M)$, $\text{rk}_B(T, M)$ as introduced in Notation 15.2.

Theorem 1.4 (Corollary 16.4, Proposition 16.5). *The following always holds:*

$$\delta(M) = \delta_{\text{Ta}}(M) + \delta_2(M) + \delta_{\text{alg}}(M) + \delta_3(M) + \delta_{\text{inc2}}(M) + \delta_{\text{inc3}}(M).$$

1. *All Tate periods are $\overline{\mathbb{Q}}$-multiples of $2\pi i$. All algebraic periods are in $\overline{\mathbb{Q}}$. In particular, $\delta_{\text{Ta}}(M)$ and $\delta_{\text{alg}}(M)$ take the values 0 or 1, depending on the (non)-vanishing of T and L.*
2. *We have*

$$\delta_2(M) = \sum_B \frac{4g(B)^2}{e(B)},$$

   where the sum is taken over all simple factors of A, without multiplicities.
3. *We have*

$$\delta_3(M) = \sum_B 2g(B)\text{rk}_B(L, M),$$

$$\delta_{\text{inc2}}(M) = \sum_B 2g(B)\text{rk}_B(T, M).$$

The special case $A = 0$ gives Baker's Theorem. The most interesting and hardest contribution is $\mathcal{P}_{\text{inc3}}(M)$. The computation of this contribution was not possible without the methods that we develop here. Up to particular cases the formulas for the other contributions were not in the literature either. For an overview see, for example [BW07, Section 6.2], [Wüs84a, Wüs12] and [Wüs21].

The formula for $\mathcal{P}_{\text{inc3}}(M)$ simplifies in the case of motives that we call saturated; see Definition 15.1.

Theorem 1.5 (Theorem 15.3). *If $M = M_0 \times M_1$ is the product of a Baker motive $M_0 = [L_0 \to T_0]$, i.e. with vanishing abelian part, and a saturated motive $M_1 = [L_1 \to G_1]$, then*

$$\delta_{\text{inc3}}(M) = \text{rk}_{\text{gm}}(L, M_1) + \sum_B e(B)\text{rk}_B(G_1, M_1)\text{rk}_B(L_1, M_1).$$

Fortunately, by Theorem 15.3 (2) the periods of a general motive are always included in the period space of $M_0 \times M_{\text{sat}}$ with M_0 of Baker type ($A_0 = 0$) and M_{sat} saturated.

There is a precise recipe for $\delta_{\text{inc3}}(M)$ for any 1-motive M. It is spelt out in Chapter 17, in particular Theorem 17.8. See also Chapter 11 for examples of elliptic curves without and with CM.

1.4 Method of Proof

As in the case of closed paths, the main ingredient of our proof (and the only input from transcendence theory) is the Analytic Subgroup Theorem of [Wüs89]. We give a reformulation as Theorem 6.2: given a smooth connected commutative algebraic group over $\overline{\mathbb{Q}}$ and $u \in \mathrm{Lie}(G^{\mathrm{an}})$ such that $\exp_G(u) \in G(\overline{\mathbb{Q}})$, there is a canonical short exact sequence

$$0 \to G_1 \to G \xrightarrow{\pi} G_2 \to 0$$

of algebraic groups over $\overline{\mathbb{Q}}$ such that $\mathrm{Ann}(u) = \pi^*(\mathrm{coLie}(G_2))$ and $u \in \mathrm{Lie}(G_1^{\mathrm{an}})$. Here $\mathrm{Ann}(u) \subset \mathrm{coLie}(G)$ is the largest subspace such that $\langle \mathrm{Ann}(u), u \rangle = 0$ under the canonical pairing.

Given a 1-motive M, Deligne constructed a vector extension $M^{\natural}$ of G such that $V_{\mathrm{dR}}(M) = \mathrm{Lie}(M^{\natural})$. This is the group to which we apply the Subgroup Theorem.

Theorem 1.6 (Subgroup Theorem for 1-motives, Theorem 9.7). *Given a 1-motive M over $\overline{\mathbb{Q}}$ and $u \in V_{\mathrm{sing}}(M)$, there is a short exact sequence of 1-motives up to isogeny*

$$0 \to M_1 \xrightarrow{i} M \xrightarrow{p} M_2 \to 0,$$

such that $\mathrm{Ann}(u) = p^* V_{\mathrm{dR}}^{\vee}(M_2)$ *and* $u \in i_* V_{\mathrm{sing}}(M_1)$. *Here* $\mathrm{Ann}(u) \subset V_{\mathrm{dR}}^{\vee}(M)$ *is the left kernel under the period pairing. The sequence is uniquely determined by these properties.*

Given a pair of non-zero $u \in V_{\mathrm{sing}}(M)$ and $\omega \in V_{\mathrm{dR}}^{\vee}(M)$ with vanishing period, the theorem provides a proper submotive M_1 of M such that $u = i_* u_1$ for $u_1 \in V_{\mathrm{sing}}(M_1)$ and $\omega = p^* \omega_2$ for $\omega_2 \in V_{\mathrm{dR}}^{\vee}(M_2)$. Any $\overline{\mathbb{Q}}$-linear relation between periods can be translated into the vanishing of a period. Then the Subgroup Theorem for 1-motives is applied.

As a by-product, we also get a couple of new results on 1-motives over $\overline{\mathbb{Q}}$: they are a full subcategory of the category of $\mathbb{Q}$-Hodge structures over $\overline{\mathbb{Q}}$ (see Proposition 8.17) and of the category of (non-effective) Nori motives (see Theorem 13.5) and of the category $(\mathbb{Q}, \overline{\mathbb{Q}})$-Vect of pairs of vector spaces together with a period matrix. The last statement was also obtained independently by Andreatta, Barbieri-Viale and Bertapelle; see [ABVB20]. The case of Hodge structures has just recently been considered by André in [And21]. He proves that the functor from 1-motives into $\mathbb{Q}$-Hodge structures is fully faithful for all algebraically closed fields $k \subset \mathbb{C}$.

1.5 Why 1-Motives?

This seems the right moment to address the question of whether our emphasis on 1-motives is necessary. We think that the answer is yes.

Obviously, all proofs using 1-motives could be rewritten in terms of commutative algebraic groups because this is how the Subgroup Theorem for 1-Motives itself is deduced. However, the dimension formulas depend on the constituents of the 1-motive and do not admit a transparent formulation in terms of the constituents of the algebraic group.

More generally, 1-motives are the link between the classical objects of transcendence theory à la Lindemann, Schneider or Baker and the structural predictions linked with Grothendieck, André or Kontsevich.

1.6 The Case of Elliptic Curves

The above results are very general and depend on a subtle interplay between the data. It is a non-trivial task to make them explicit in particular examples. We have carried this out to some extent in the case of an elliptic curve E defined over $\overline{\mathbb{Q}}$.

Recall the Weierstraß $\wp$-, ζ- and σ-functions for E. We obtain the following result.

Theorem 1.7 (Theorem 18.6). *Let $u \in \mathbb{C}$ be such that $\wp(u) \in \overline{\mathbb{Q}}$ and $\exp_E(u)$ is non-torsion in $E(\overline{\mathbb{Q}})$. Then*

$$u\zeta(u) - 2\log\sigma(u)$$

is transcendental.

This is an incomplete period integral of the third kind. The proof of the above result is actually not a direct consequence of Theorem 1.1 but rather uses the insights of our dimension computations.

We also carry out the dimension computation in this case: let $M = [L \to G]$ with $L \cong \mathbb{Z}$, G an extension of E by $\mathbb{G}_m$ that is non-split up to isogeny, $L_\mathbb{Q} \to E_\mathbb{Q}$ injective. Then by Propositions 11.1 and 11.3,

$$\dim \mathcal{P}\langle M \rangle = \begin{cases} 11 & E \text{ without CM,} \\ 9 & E \text{ CM.} \end{cases}$$

The incomplete periods of the third kind become more difficult already if we consider $M = [L \to G]$ with $L \cong \mathbb{Z}^2$, G an extension of E by $\mathbb{G}_m^2$, again $L_\mathbb{Q} \to E_\mathbb{Q}$ injective and G completely non-split up to isogeny. If E does not have CM, then

$$\dim \mathcal{P}\langle M \rangle = 18.$$

If E is CM, then

$$\dim \mathcal{P}\langle M \rangle = 16, 14, 12, 10,$$

depending on the interplay of the complex multiplication and L and G. The extreme case occurs when $\mathrm{End}(M)$ is the CM-field. Then the resulting dimension is 10.

1.7 Values of Hypergeometric Functions

Euler had already known that the hypergeometric function $F(a, b, c; z)$ can be written as a quotient of two integrals. If a, b, c are rational numbers, these integrals can be regarded as periods on certain explicit algebraic curves. Knowledge about $\overline{\mathbb{Q}}$-linear indepdendence of periods then translates into transcendence statements for the values $F(a, b, c; \lambda)$ for $\lambda \in \overline{\mathbb{Q}} \smallsetminus \{0, 1\}$. This insight is exploited by Wolfart in [Wol88] and by Chudnovsky–Chudnovsky in [CC88]. We explain the argument in detail for $a = b = 1/2$ and $c = 1$:

Proposition 1.8 (Proposition 19.3). *The value $F(1/2, 1/2, 1; z)$ of the hypergeometric function is transcendental for $z \in \overline{\mathbb{Q}} \smallsetminus \{0, 1\}$.*

The proposition follows from the $\overline{\mathbb{Q}}$-linear independence of π and the complete periods of elliptic curves established first by Schneider in 1936; see [Sch37, Satz IIIa].

In the case of general $a, b, c \in \mathbb{Q}$ with least common denominator N, the Euler integrals can be seen as periods for the algebraic curve with affine equation

$$y^N = x^r (1 - x)^s (1 - \lambda x)^t$$

for suitable r, s, t. For the formula in the case of $N = p$ a prime, see Proposition 19.19. These curves have been intensely studied. Using results of Gross–Rohrlich [GR78], Archinard [Arc03b] and Asakura–Otsubo [AO18], we work out another example.

Theorem 1.9 (Corollary 19.22). *Let p be a prime such that $p \not\equiv 1 \mod 3$, $1 \le n \le p - 1$. Let $0 < r, s < p$ such that p does not divide $r + s$, put $t = p - s$ and*

$$u = \left[\frac{nr}{p} \right], \quad v = \left[\frac{ns}{p} \right], \quad w = \left[\frac{nt}{p} \right].$$

We further assume

$$\left\langle \frac{nr}{p} \right\rangle + \left\langle \frac{ns}{p} \right\rangle - \left\langle \frac{n(r+s)}{p} \right\rangle \neq 1.$$

Then, for all $\lambda \in \overline{\mathbb{Q}} \setminus \{0,1\}$, *the corresponding value* $F(a,b,c;\lambda)$ *is zero or transcendental and transcendental if* $\lambda \in (0,1)$.

An explict example where the assumptions are satisfied is $p = 11$, $r = s = 2$, $n = 1, 2, 6, 7, 8$. We deduce, for example, that the numbers $F(6/11, 6/11, 12/11; \lambda)$ are zero or transcendental, provided $\lambda \in \overline{\mathbb{Q}} \setminus \{0,1\}$.

We should stress that this application relies only on complete periods on abelian varieties and not on the more general theory developed in our monograph. It should be seen as a proof of concept: the same argument can be applied to other geometric families of curves, allowing families of differential forms of the third kind and non-closed paths. The hyergeometric function would be replaced by the solutions of differential equations defined by the Gauss–Manin connection.

1.8 Structure of the Monograph

We have tried to make the monograph accessible to readers who are not familiar with either motives or periods.

The first part provides foundational material that will be used throughout, for example terminology from category theory, a review of the theory of the generalised Jacobian and the basics on singular homology and de Rham cohomology. We provide precise references for the facts that we need later. Along the way we also fix notation and normalisations. Depending on their background, readers are invited to skip some or all of these chapters and use them only for reference.

Chapters 6 and 7 address less classical material. The first deduces a reformulation of our key tool, the Analytic Subgroup Theorem. We apply it to the comparison between analytic and algebraic homomorphisms between connected commutative algebraic groups.

Chapter 7 presents an abstract formulation of the theory of periods and the Period Conjecture for abelian categories without a tensor structure.

Part II is the heart of the monograph and presents our main result. It addresses periods of 1-motives. After settling some notation, Chapter 8 starts by reviewing Deligne's category of 1-motives and its properties. We then establish auxiliary results that are needed in the next chapter.

Chapter 9 discusses periods of 1-motives and proves the version of the Period Conjecture purely in terms of 1-motives. We then consider examples: in Chapter 10 we treat the classical cases like the transcendence of π and values of log in our language. In Chapter 11 we apply the general results in the case of a 1-motive whose constituents are as small as possible without being trivial and compute the dimensions of their period spaces.

In Part III we turn to periods of algebraic varieties. Chapter 12 clarifies the notion of a cohomological period. After defining $\mathcal{P}^1$ in a down-to-earth way, the interpretation of cohomological periods as the periods of 1-motives is explained. Finally, we explain the interpretation as periods of Nori or Voevodsky motives.

In Chapter 13 we use the results on periods of 1-motives to deduce the qualitative results on $\mathcal{P}^1$ and periods of curves: the criterion on transcendence and the Period Conjecture. The results are made explicit in the classical terms of differential forms of the first, second and third kind on an algebraic curve in Chapter 14.

Part IV aims at a dimension formula for the space of periods of a 1-motive in terms of its data. Chapter 15 treats mainly the saturated case. This can be applied to deduce complete structural results in Chapter 16. Finally, Chapter 17 is devoted to an explicit dimension computation for the space of incomplete periods of the third kind, which is very involved. In this rather complicated case the results were unexpected.

In Chapter 18 we deal with the case of elliptic curves and make our results explicit in terms of the classical Weierstraß functions $\wp, \zeta, \sigma$.

We explain in Chapter 19 how transcendence results on special values of the hypergeometric function can be deduced from $\overline{\mathbb{Q}}$-linear independence of 1-periods.

There are three appendices: the first two sketch the theories of Nori and Voevodsky motives to the extent used in the proof of Theorem 13.3.

The last appendix is of a technical nature: we need to verify that the singular and de Rham realisations of a 1-motive agree with the realisation of the attached geometric motive.

PART ONE

FOUNDATIONS

2

Basics on Categories

In this monograph some basic concepts of the theory of categories are used frequently. For the convenience of the reader, we recall them here. At some places later in the text, however, the requirements are higher than those we provide in this chapter. In each case we give a precise reference to the literature where the reader can find the full information needed.

2.1 Additive and Abelian Categories

Definition 2.1. A *category* C consists of a class of *objects* $\mathrm{Ob}(C)$ and for every pair of objects $X, Y \in \mathrm{Ob}(C)$ a set of *morphisms* $\mathrm{Mor}_C(X, Y)$ together with a composition law,

$$\circ \colon \mathrm{Mor}_C(X, Y) \times \mathrm{Mor}_C(Y, Z) \to \mathrm{Mor}_C(X, Z),$$
$$(f, g) \qquad \mapsto g \circ f,$$

for every triple of objects X, Y, Z. These data are subject to the following conditions:

1. The composition law is associative.
2. For every object X there is a morphism $\mathrm{id}_X \in \mathrm{Mor}_C(X, X)$ which is a left and right identity for the composition law, i.e. $f \circ \mathrm{id}_X = f$ for all Y and all $f \colon X \to Y$ and $\mathrm{id}_X \circ g = g$ for all Z and all $g \colon Z \to X$.

A category is $\mathbb{Z}$-linear (or $\mathbb{Q}$-linear) if the morphism sets are given the structure of abelian groups (or $\mathbb{Q}$-vector spaces) and the composition of morphisms is bilinear. We usually write Hom_C instead of Mor_C in this case. It is *additive* if it is $\mathbb{Z}$-linear and, in addition, has finite direct sums. The direct sum of two objects X and Y is denoted $X \oplus Y$. As the particular case of the empty direct sum, this also requires the existence of a 0-object characterised uniquely by the property that

$$\mathrm{Hom}(0, X) = \mathrm{Hom}(X, 0) = 0$$

for all objects X.

An additive category $\mathcal{A}$ is called *abelian* if the following two properties are satisfied:

1. Every morphism $f\colon X \to Y$ in $\mathcal{A}$ has a kernel and cokernel.
2. For every morphism $f\colon X \to Y$ in $\mathcal{A}$ the natural map $X/\mathrm{ker}(f) \to \mathrm{im}(f)$ is an isomorphism.

In this abstract setting, the image of a morphism is defined as the kernel of the cokernel.

Example 2.2. For every ring R, the category of R-modules is abelian. The category of finitely generated modules is additive. It is abelian if R is noetherian.

Many examples of additive and abelian categories are going to be used throughout the book. In order of appearance:

Examples 2.3. 1. Let k be a field. Then the category of connected commutative algebraic groups schemes over k is additive, but not abelian. We refer the reader to Chapter 4 for details.
2. Let $K, L \subset \mathbb{C}$ be subfields. Then the category (K, L)-Vect introduced in Section 7.2 is $\mathbb{Q}$-linear and abelian.
3. Let k be a field. Then the category of filtered k-vector spaces is additive but not abelian. Every morphism has a kernel and a cokernel, but the isomorphism between $X/\mathrm{ker}(f)$ and $\mathrm{im}(f)$ fails in general.
4. Let k be an algebraically closed field of characteristic 0. The category 1-Mot$_k$ of iso-1-motives is abelian. A thorough review is given in Chapter 8.

Given a $\mathbb{Z}$-linear category $\mathcal{A}$, we obtain a $\mathbb{Q}$-linear category $\mathcal{A} \otimes \mathbb{Q}$ with the same objects as $\mathcal{A}$ and morphism

$$\mathrm{Hom}_{\mathcal{A} \otimes \mathbb{Q}}(X, Y) = \mathrm{Hom}_{\mathcal{A}}(X, Y) \otimes_{\mathbb{Z}} \mathbb{Q}.$$

We refer to it as the *isogeny category of $\mathcal{A}$*. If $\mathcal{A}$ is additive or abelian, then so is $\mathcal{A} \otimes \mathbb{Q}$.

Examples 2.4. 1. Let k be a field. If $\mathcal{A}$ is the category of abelian varieties over k, then $\mathcal{A} \otimes \mathbb{Q}$ is what is referred to as the *category of abelian varieties up to isogeny* in the literature. It is abelian. The same remark also applies to the category of connected commutative group schemes.

2. By Definition 8.1, 1-Mot$_k$ is defined as the isogeny category of the category of 1-motives over k.

For any category C, we define the *additive hull* $\mathbb{Z}[C]$, where the objects are formal direct sums $\bigoplus_{i=1}^{n} X_i$ for $n \geq 0$ and $X_1, \ldots, X_n \in C$. We interpret the empty direct sum as an object 0. Morphisms are defined by the formula

$$\mathrm{Hom}_{\mathbb{Z}[C]}\left(\bigoplus_{i=1}^{n} X_i, \bigoplus_{j=1}^{m} Y_j \right) = \bigoplus_{i,j} \mathbb{Z}[\mathrm{Hom}_C(X_i, Y_j)].$$

Here for a set S, we denote by $\mathbb{Z}[S]$ the free abelian group with basis S.

2.2 Subcategories

Given a category C, a *subcategory* of C is a category C' such that every object and every morphism of C' is an object and morphism in C, respectively. The composition of morphisms in C' is defined as their composition in C and the identity morphisms in C' agree with the identity morphisms in C. A subcategory is called *full* if

$$\mathrm{Mor}_{C'}(X, Y) = \mathrm{Mor}_C(X, Y)$$

for all objects X, Y of C'.

Remark 2.5. If C is additive or abelian, then a subcategory C' is not necessarily additive or abelian itself. If C and C' are both abelian, this does not imply that the kernels and cokernels are the same when computed in C or C'. We are not going to consider such pathological situations, which only appear if the subcategory is not full.

Example 2.6. The category of abelian varieties (up to isogeny) is a full subcategory of the category of connected commutative algebraic groups (up to isogeny).

Example 2.7. The category of $\mathbb{Q}$-vector spaces is a full subcategory of the category of abelian groups.

Let $\mathcal{A}$ be an abelian category. A *subquotient* of an object X in $\mathcal{A}$ is a quotient of a subobject of X, or equivalently, a subobject of a quotient of X.

Definition 2.8. Let $\mathcal{A}$ be an abelian category and $X \in \mathcal{A}$ an object. We define $\langle X \rangle$ as the smallest full subcategory closed under subquotients containing X.

More explicitly, this means that $\langle X \rangle$ contains X and all the quotients and subobjects of every object Y of $\langle X \rangle$.

Lemma 2.9. *The category $\langle X \rangle$ is abelian. We have*

$$\mathcal{A} = \bigcup_{X \in C} \langle X \rangle.$$

Proof Let $f \colon Y \to Z$ be a morphism in $\langle X \rangle$. Then $\ker(f) \subset Y$ exists in $\mathcal{A}$. As a subobject of an object in $\langle X \rangle$, it is itself an object of $\langle X \rangle$. The universal property of a kernel holds because it holds in $\mathcal{A}$. The same argument gives the existence of cokernels. The natural map $Y / \ker(f) \to \mathrm{im}(f)$ has an inverse in $\mathcal{A}$ because the category is abelian. This inverse is in $\langle X \rangle$ because the subcategory is full.

Obviously all objects of $\mathcal{A}$ are contained in the union of all $\langle X \rangle$. We have to check that the same is true for morphisms. Let $f \colon X \to Y$ be a morphism in $\mathcal{A}$. Both X and Y are subobjects of $X \oplus Y$, hence they are both objects of $\langle X \oplus Y \rangle$. As $\langle X \oplus Y \rangle \subset \mathcal{A}$ is a full subcategory, the morphism f is a morphism in $\langle X \oplus Y \rangle$. $\square$

2.3 Functors

Definition 2.10. Let C and C' be categories. A *covariant functor* $F \colon C \to C'$ is an assignment $F \colon \mathrm{Ob}(C) \to \mathrm{Ob}(C')$ together with a map

$$\mathrm{Mor}_C(X, Y) \to \mathrm{Mor}_{C'}(F(X), F(Y))$$

for every pair of objects X, Y of C. It is subject to the following conditions.

1. Compatibility with composition: $F(g) \circ F(f) = F(g \circ f)$ for all objects X, Y, Z in C and morphisms $f \colon X \to Y$, $g \colon Y \to Z$.
2. Compatibility with identities: $F(\mathrm{id}_X) = \mathrm{id}_{F(X)}$ for all objects X of C.

In the case of a *contravariant functor*, we are given maps

$$\mathrm{Mor}_C(X, Y) \to \mathrm{Mor}_{C'}(F(Y), F(X))$$

and the compatibility condition reads $F(f) \circ F(g) = F(g \circ f)$ instead.

A functor $F \colon C \to C'$ is called *faithful, full* or *fully faithful* if for all objects $X, Y \in C$ the natural map

$$\mathrm{Hom}_C(X, Y) \to \mathrm{Hom}_{C'}(F(X), F(Y))$$

is injective, surjective or bijective, respectively.

Example 2.11. If F is the inclusion of a subcategory of a category, then it is faithful. If the subcategory is full, the inclusion is fully faithful.

A functor $F: \mathcal{C} \to \mathcal{C}'$ between $\mathbb{Z}$-linear or $\mathbb{Q}$-linear categories is called *additive* or $\mathbb{Q}$-linear if for all $X, Y \in \mathcal{C}$ the map

$$\mathrm{Hom}_{\mathcal{C}}(X, Y) \to \mathrm{Hom}_{\mathcal{C}'}(F(X), F(Y))$$

is $\mathbb{Z}$-linear or $\mathbb{Q}$-linear, respectively. Such a functor automatically respects direct sums, provided that they exist (e.g. because the categories are additive).

An additive functor $F: \mathcal{A} \to \mathcal{A}'$ between abelian categories is called *exact* if it sends short exact sequences to short exact sequences.

Lemma 2.12. *Let $F: \mathcal{A} \to \mathcal{A}'$ be an exact functor between abelian categories. Then F is faithful if and only if for all X in $\mathcal{A}$ the assumption $F(X) = 0$ implies $X \cong 0$.*

Proof Assume that F is faithful and that X is an object of $\mathcal{A}$ such that $F(X) = 0$. Then this implies $F(0) = F(\mathrm{id}_X)$. By faithfulness this gives $0 = \mathrm{id}_X$ and hence $X \cong 0$.

Conversely assume the condition on objects. Let f be in the kernel of the map $\mathrm{Hom}_{\mathcal{A}}(X, Y) \to \mathrm{Hom}_{\mathcal{A}'}(F(X), F(Y))$. The functor F maps the exact sequence

$$0 \to \ker(f) \to X \xrightarrow{f} Y \to \mathrm{coker}(f) \to 0$$

to the exact sequence

$$0 \to F(\ker(f)) \to F(X) \xrightarrow{F(f)=0} F(Y) \to F(\mathrm{coker}(f)) \to 0.$$

This gives $F(\ker(f)) \cong F(X)$ and $F(Y) \cong F(\mathrm{coker}(f))$. Now consider the short exact sequence

$$0 \to \ker(f) \to X \to X/\ker(f) \to 0$$

and its image

$$0 \to F(\ker(f)) \to F(X) \to F(X/\ker(f)) \to 0.$$

We had established that the first map is an isomorphism, so $F(X/\ker(f)) \cong 0$. By assumption this implies that $X/\ker(f) \cong 0$ or $\ker(f) \cong X$. The same type of argument also shows that $Y \cong \mathrm{coker}(f)$. Taken together this means that $f = 0$. □

Faithful functors allow us to test for inclusions.

Lemma 2.13. *Let $F: \mathcal{A} \to \mathcal{A}'$ be a faithful exact functor between abelian categories, $X \in \mathcal{A}$ an object and $X_1, X_2 \subset X$ subobjects. If $F(X_2) \subset F(X_1)$, then $X_2 \subset X_1$.*

Proof Let $X_3 = X_1 \cap X_2$ (or, more abstractly, let X_3 be the pull-back of $X_1 \to X$ via $X_2 \to X$). We need to show that the natural inclusion $X_3 \to X_2$ is an isomorphism, whence $X_2 \subset X_1$. By the exactness of F, we have $F(X_3) \cong F(X_1) \cap F(X_2)$. By assumption this is $F(X_2)$. We apply F to the exact sequence

$$0 \to X_3 \to X_2 \to C \to 0.$$

As $F(X_3) = F(X_2)$, we get $F(C) = 0$. By the faithfulness of F, this implies $C \cong 0$. □

As a consequence of our results on transcendence and the Period Conjecture, we are are also going to establish results on fullness of certain functors; see Proposition 8.17, Theorem 9.14 and Theorem 13.5. The following criterion will be useful.

Lemma 2.14. *Let $F \colon \mathcal{A} \to \mathcal{A}'$ be a faithful exact functor between abelian categories. Assume that the image of F is closed under subquotients, i.e. if*

$$0 \to Y' \to F(X) \to Y'' \to 0$$

is an exact sequence in $\mathcal{A}'$, then there is a short exact sequence

$$0 \to X' \to X \to X'' \to 0$$

in $\mathcal{A}$ mapping to the given exact sequence in $\mathcal{A}'$. Then F is full.

Proof Let $f \colon F(Y_1) \to F(Y_2)$ be a morphism in $\mathcal{A}'$ and $\Gamma \subset F(Y_1) \times F(Y_2)$ its graph. We find the graph as the image of $F(Y_1)$ under the map

$$F(Y_1) \xrightarrow{\Delta} F(Y_1) \times F(Y_1) \xrightarrow{(\mathrm{id}, f)} F(Y_1) \oplus F(Y_2).$$

It is a subobject. By assumption, there is $G \subset Y_1 \times Y_2$ in $\mathcal{A}$ such that $F(G) = \Gamma$. The projection $p \colon G \to Y_1 \times Y_2 \to Y_1$ is an isomorphism because this is true for the image $\Gamma \to F(Y_1)$ and F is faithful. Let i be its inverse. The composition

$$Y_1 \xrightarrow{i} G \subset Y_1 \times Y_2 \to Y_2$$

is the preimage of f we were looking for. □

3

Homology and Cohomology

A key step in our approach to periods is the reinterpretation of paths as homology classes and differential forms as classes in algebraic de Rham cohomology. We survey the key definitions. For an in-depth review with complete references, we refer the reader to [HMS17, Part I].

3.1 Singular Homology

All topological spaces in this section are locally compact, Hausdorff and satisfy the second countable axiom. The analytification of an algebraic variety over $\mathbb{C}$ is an example of such a space. We refer to standard textbooks on algebraic topology like [Spa66] or [Hat02] for full details and proofs.

Definition 3.1. The *topological n-simplex* Δ_n is defined by

$$\Delta_n = \left\{ (x_0, \ldots, x_n) \in \mathbb{R}^{n+1} \mid x_0 \geq 0, \ldots, x_n \geq 0, \sum_{i=0}^{n} x_i = 1 \right\}.$$

By setting one coordinate x_i to 0, we get the inclusion of the codimension 1 *faces F_i*. They are homeomorphic to the $(n-1)$-simplex in an obvious way.

Example 3.2. The 0-simplex is a single point. The 1-simplex is the interval from $F_0 = (0, 1)$ to $F_1 = (1, 0)$. We often identify it with the unit interval $[0, 1] \subset \mathbb{R}$.

Definition 3.3. Let X be a topological space. For an integer $n \geq 0$ a *singular n-chain* is a formal $\mathbb{Q}$-linear combination of continuous maps $f \colon \Delta_n \to X$. The space of singular chains $S_n(X)$ is a $\mathbb{Q}$-vector space. Together with the natural differential

$$d_n \colon S_n(X) \to S_{n-1}(X),$$

which maps a basis element f in $S_n(X)$ to

$$d_n(f) = \sum_{i=0}^{n}(-1)^i f|_{F_i},$$

we obtain a chain complex $(S_*(X), d_*)$, the *singular chain complex of X*. Its homology

$$H_n^{\mathrm{sing}}(X, \mathbb{Q}) := H_n((S_*(X), d_*))$$

is called the *singular homology of X with rational coefficients*.

Example 3.4. If X is a finite discrete set, then

$$H_n^{\mathrm{sing}}(X, \mathbb{Q}) = \begin{cases} \mathbb{Q}^{|X|} & n = 0, \\ 0 & \text{otherwise.} \end{cases}$$

Example 3.5. Take $\sigma = \sum_{i=1}^{n} a_i\gamma_i \in S_1(X)$, where $a_i \in \mathbb{Q}$ and $\gamma_i\colon \Delta_1 \to X$ is continuous. We identify Δ_1 with the unit interval and view the γ_i as paths. Then σ is in the kernel of d_1 if the formal linear combination

$$\sum_{i=1}^{n} a_i\gamma_i(1) - \sum_{i=1}^{n} a_i\gamma_i(0)$$

vanishes. We say that σ is *closed* or a *cycle*. Cycles in the image of d_2 are called *boundaries*. They are the ones that can be filled in by discs.

In particular, every closed path gives rise to a homology class, and homotopic paths are homologous. We get a well-defined map

$$\pi_1(X, x_0) \to H_1^{\mathrm{sing}}(X, \mathbb{Q}),$$

the *Hurewicz homomorphism*. Its image generates $H_1^{\mathrm{sing}}(X, \mathbb{Q})$ if X is path connected. In fact, we have

$$H_1^{\mathrm{sing}}(X, \mathbb{Q}) \cong \pi_1(X, x_0)^{\mathrm{ab}} \otimes_{\mathbb{Z}} \mathbb{Q}$$

in this case.

Example 3.6. Let C be a smooth complete curve of genus g over $\mathbb{C}$ and C^{an} the compact Riemann surface defined by its analytification. Then

$$\dim_{\mathbb{Q}} H_1^{\mathrm{sing}}(C^{\mathrm{an}}, \mathbb{Q}) = 2g.$$

For $C^\circ = C \smallsetminus S$ with S a non-empty finite set, we have

$$\dim_{\mathbb{Q}} H_1^{\mathrm{sing}}(C^{\circ\mathrm{an}}, \mathbb{Q}) = 2g - |S| + 1.$$

Example 3.7. The projective line satisfies

$$H_1^{\text{sing}}(\mathbb{P}^{1\,\text{an}}, \mathbb{Q}) = 0.$$

For $\mathbb{A}^{1\,\text{an}} = \mathbb{C}$ and $\mathbb{G}_m^{\text{an}} = \mathbb{C}^*$ we obtain

$$H_1^{\text{sing}}(\mathbb{C}, \mathbb{Q}) = 0, \quad H_1^{\text{sing}}(\mathbb{C}^*, \mathbb{Q}) \cong \mathbb{Q},$$

with the last group generated by a loop around 0, for example, the boundary of the unit disc.

We also want to handle non-closed paths. They define classes in relative homology.

Definition 3.8. Let X be a topological space, $A \subset X$ a closed subset. We call

$$S_*(X, A; \mathbb{Q}) = S_*(X, \mathbb{Q})/S_*(A, \mathbb{Q})$$

the *singular chain complex for* (X, A). Its homology is *singular homology of the pair* (X, A) *with rational coefficients* or *singular homology of X relative to A, written as*

$$H_n^{\text{sing}}(X, A; \mathbb{Q}) = H_n(S_*(X, A; \mathbb{Q}), d_*).$$

By definition, the relative singular homology is functorial for pairs.

Example 3.9. Let $\gamma \colon [0, 1] \to X$ be a path with endpoints in A. Then it is a cycle relative to A and gives rise to a class in $H_1^{\text{sing}}(X, A; \mathbb{Q})$.

The short exact sequence

$$0 \to S_*(A, \mathbb{Q}) \to S_*(X, \mathbb{Q}) \to S_*(X, A; \mathbb{Q}) \to 0$$

gives rise to a long exact sequence in homology

$$\cdots \to H_n^{\text{sing}}(A, \mathbb{Q}) \to H_n^{\text{sing}}(X, \mathbb{Q}) \to H_n^{\text{sing}}(X, A; \mathbb{Q}) \to H_{n-1}^{\text{sing}}(A, \mathbb{Q}) \to \cdots.$$

Of particular interest for us is $n = 1$.

Example 3.10. If A is a finite set of points, then the sequence simplifies to

$$0 \to H_1^{\text{sing}}(X, \mathbb{Q}) \to H_1^{\text{sing}}(X, A; \mathbb{Q}) \to \mathbb{Q}^{|X|-1} \to 0.$$

More generally, we get *boundary maps* for triples $B \subset A \subset X$

$$\partial \colon H_n^{\text{sing}}(X, A; \mathbb{Q}) \to H_{n-1}^{\text{sing}}(A, B; \mathbb{Q})$$

and natural long exact sequences

$$\cdots \to H_n^{\text{sing}}(A, B; \mathbb{Q}) \to H_n^{\text{sing}}(X, B; \mathbb{Q}) \to H_n^{\text{sing}}(X, A; \mathbb{Q}) \to H_{n-1}^{\text{sing}}(A, B; \mathbb{Q}) \to \cdots.$$

Remark 3.11. If X is a manifold, it suffices to work with *smooth singular chains*. Let $S_*^\infty(X) \subset S_*(X)$ be the subcomplex of linear combinations of C^∞-maps $f: \Delta_n \to X$; see [HMS17, Definition 2.2.2]. By [HMS17, Theorem 2.2.5], the complex $S_*^\infty(X)$ can be used to compute singular homology of X.

There is another tool that allows us to reduce questions on the homology of algebraic varieties to the smooth case. We formulate it in cohomology obtained by replacing all vector spaces by their duals.

Proposition 3.12 (Blow-up sequence). *Let X be an algebraic variety over $\mathbb{C}$, $\pi: \widetilde{X} \to X$ a proper map, $Z \subset X$ a closed subvariety with preimage E in $\widetilde{X}$ such that π induces an isomorphism $\widetilde{X} \smallsetminus E \to X \smallsetminus Z$. Then there is a natural long exact sequence*

$$\cdots \to H^n_{\mathrm{sing}}(X^{\mathrm{an}}, \mathbb{Q}) \to H^n_{\mathrm{sing}}(\widetilde{X}^{\mathrm{an}}, \mathbb{Q}) \oplus H^n_{\mathrm{sing}}(Z^{\mathrm{an}}, \mathbb{Q}) \to H^n_{\mathrm{sing}}(E^{\mathrm{an}}, \mathbb{Q})$$
$$\to H^{n+1}_{\mathrm{sing}}(X^{\mathrm{an}}, \mathbb{Q}) \to \cdots.$$

Proof In the case of analytic spaces attached to algebraic varieties, we can identify singular cohomology with sheaf cohomology. The statement then follows from proper base change (see [KS90, Proposition 2.6.7]) for π. $\quad\square$

Example 3.13. Let C be a curve with normalisation $\widetilde{C}$. Let $Y \subset C$ be the set of singular points with preimage $\widetilde{Y} \subset \widetilde{C}$. The assumptions of the proposition are satisfied and the long exact sequence degenerates to the short exact sequence

$$0 \to \mathbb{Q}^N \to H^1_{\mathrm{sing}}(C^{\mathrm{an}}, \mathbb{Q}) \to H^1_{\mathrm{sing}}(\widetilde{C}^{\mathrm{an}}, \mathbb{Q}) \to 0,$$

with $N = |\widetilde{Y}| - |Y|$.

3.2 Algebraic de Rham Cohomology

Algebraic de Rham cohomology was introduced by Grothendieck in [Gro66] as a purely algebraic way to define the Betti numbers of algebraic varieties. In this section k is a field of characteristic 0.

3.2.1 The Smooth Case

We start by presenting the much easier smooth case.

Definition 3.14. Let X be a smooth variety over k. We define *algebraic de Rham cohomology of X* as hypercohomology of the complex of sheaves of differential forms on X,

$$H^n_{\mathrm{dR}}(X) = \mathbb{H}^n(X, \Omega_X^*).$$

This is particularly easy if X is, in addition, affine. In this case,

$$H_{\mathrm{dR}}^n(X) = H^n(\Omega^*(X), d).$$

There are different ways to compute hypercohomology. We make the approach via Čech-cohomology explicit for later use.

Let X be a smooth variety and $\mathfrak{U} = (U_1, \ldots, U_n)$ be an open cover of X by affine subvarieties U_i. For every $I \subset \{1, \ldots, n\}$ we put $U_I = \bigcap_{i \in I} U_i$ and for every $p, q \in \mathbb{N}_0$ we define

$$C^p(\mathfrak{U}, \Omega_X^q) = \prod_{|I| = p+1} \Omega^q(U_I).$$

The $C^p(\mathfrak{U}, \Omega_X^q)$ form a double complex. The differential d in q-direction is induced by the differential of Ω_X^*. The differential δ in p-direction is the differential of the Čech-complex given as follows: for $\alpha = (\alpha_I) \in C^p(\mathfrak{U}, \Omega_X^q)$, we have

$$\delta^p(\alpha)_{i_0 \leq i_1 \leq i_p} = \sum_{j=0}^p (-1)^p \alpha_{i_0 \leq \cdots \leq \hat{i}_j \leq \cdots i_p},$$

where $\hat{i}_j$ means that the index is omitted.

Definition 3.15. Let $R\widetilde{\Gamma}_{\mathrm{dR}}(X, \mathfrak{U})$ be the total complex of the double complex $C^p(\mathfrak{U}, \Omega^q)$ consisting of

$$R\widetilde{\Gamma}_{\mathrm{dR}}(X, \mathfrak{U})^n = \bigoplus_{p+q=n} C^p(\mathfrak{U}, \Omega^q)$$

with differential $\sum_{p+q=n}(d^p + (-1)^q \delta^q)$.

Remark 3.16. This complex is nice because it is explicit and bounded. However, the boundary depends on the choice of an ordering of $U_1, \ldots, U_n$. In consequence, these complexes have bad functorial properties, unless $f \colon Y \to X$ is affine.

Lemma 3.17. *The complex $R\widetilde{\Gamma}_{\mathrm{dR}}(X, \mathfrak{U})$ computes the algebraic de Rham cohomology of X.*

Proof We take the stupid filtration on the complex Ω_X^*, which induces a filtration on the double complex $C^p(\mathfrak{U}, \Omega^q)$. By the spectral sequence for the filtration, it suffices to consider the individual Ω^qs. They are coherent, so by [Har75, III Theorem 4.5], Čech-cohomology agrees with sheaf cohomology. $\square$

We also need relative cohomology. Again the smooth case is easier.

Definition 3.18. Let X be a smooth variety, $Y \subset X$ a smooth closed subvariety, and $\mathfrak{U}$ a finite open affine cover of X. We put

$$\widetilde{R\Gamma}_{dR}(X, Y, \mathfrak{U}) := \mathrm{cone}(\widetilde{R\Gamma}_{dR}(Y, \mathfrak{U} \cap Y) \to \widetilde{R\Gamma}_{dR}(X, \mathfrak{U}))[-1]$$

and define *algebraic de Rham cohomology of the pair* (X, Y) or *algebraic de Rham cohomology of X relative to Y* as

$$H^n_{dR}(X, Y) = H^n(\widetilde{R\Gamma}_{dR}(X, Y, \mathfrak{U})).$$

These groups satisfy the same formal properties as singular cohomology.

Lemma 3.19. *Relative algebraic de Rham cohomology is well defined. There is a natural long exact sequence*

$$\cdots \to H^n_{dR}(X, Y) \to H^n_{dR}(X) \to H^n_{dR}(Y) \to H^{n+1}_{dR}(X, Y) \to \cdots.$$

Proof The exact sequence is the long exact sequence attached to the distinguished triangle

$$\widetilde{R\Gamma}_{dR}(X, \mathfrak{U}) \to R\Gamma_{dR}(Y, \mathfrak{U} \cap Y) \to \widetilde{R\Gamma}_{dR}(X, Y, \mathfrak{U})[1]$$

or, in different language, the short exact sequence of complexes

$$0 \to \widetilde{R\Gamma}_{dR}(Y, \mathfrak{U} \cap Y)[-1] \to \widetilde{R\Gamma}_{dR}(X, Y, \mathfrak{U}) \to \widetilde{R\Gamma}_{dR}(X, \mathfrak{U}) \to 0.$$

To verify that algebraic de Rham cohomology is well defined we need to check independence of the choice of cover. We sketch the argument.

As a first step, replace $\widetilde{R\Gamma}_{dR}(X, \mathfrak{U})$ by the quasi-isomorphic complex $\widetilde{R\Gamma}_{dR}(X, \mathfrak{U})'$, which has all tuples $I = (i_0, \ldots, i_n)$ as indices rather than only the ordered ones. Given two covers $\mathfrak{U}_1$ and $\mathfrak{U}_2$, there is a common refinement $\mathfrak{U}_3$. It suffices to compare $\mathfrak{U}_1$ and $\mathfrak{U}_2$ with $\mathfrak{U}_3$. The choice of a refinement map from $\mathfrak{U}_3$ to $\mathfrak{U}_1$ induces homomorphisms

$$\widetilde{R\Gamma}_{dR}(X, \mathfrak{U}_1)' \to \widetilde{R\Gamma}_{dR}(X, \mathfrak{U}_3)'$$

and also for Y and the pair (X, Y). They are quasi-isomorphisms for X and Y because the complexes compute algebraic de Rham cohomology. By the above-mentioned long exact sequence the comparison map is a quasi-isomorphism for (X, Y) as well. $\square$

If $\dim X = 0$, then $\Omega^*_X = \mathcal{O}_X[0]$ and X is affine, hence

$$\widetilde{R\Gamma}_{dR}(X) = \mathcal{O}_X(X).$$

We now spell out the curve case in degree 1. Let C be a smooth affine curve over k, $D \subset C$ a finite set of closed points viewed as a smooth subvariety of dimension 0. We use the trivial cover $\mathfrak{U} = (C)$ and get

$$R\widetilde{\Gamma}_{\mathrm{dR}}(C, \mathfrak{U}) = [\mathcal{O}(C) \xrightarrow{f \mapsto (df, -f|_D)} \Omega^1(C) \oplus \mathcal{O}(D)] \qquad (3.1)$$

with $\mathcal{O}(C)$ in degree 0.

More generally, if C is not necessarily affine, let $U_1, \ldots, U_n$ be an open affine cover. We write $D_i = U_i \cap D$ and more generally $D_I = D \cap U_I$. By definition $R\widetilde{\Gamma}(C, D, \mathfrak{U})$ is the shifted cone of the homomorphism $R\widetilde{\Gamma}(C, \mathfrak{U}) \to R\widetilde{\Gamma}(D, \mathfrak{U} \cap D)$ of complexes. If we write the complexes vertically this takes the form

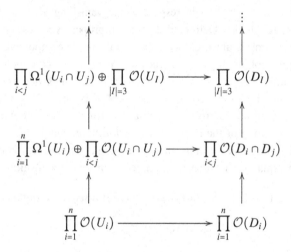

When taking the cone, we obtain the following description of cohomology in degree 1.

Lemma 3.20. *The group $H^1_{\mathrm{dR}}(C, D)$ is given by the cohomology in degree 1 of*

$$\prod_{i=1}^{n} \mathcal{O}(U_i) \longrightarrow \prod_{i=1}^{n} \Omega^1(U_i) \oplus \prod_{i<j} \mathcal{O}(U_i \cap U_j) \oplus \prod_{i=1}^{n} \mathcal{O}(D_i)$$

$$\longrightarrow \prod_{i<j} \Omega^1(U_i \cap U_j) \oplus \prod_{|I|=3} \mathcal{O}(U_I) \oplus \prod_{i<j} \mathcal{O}(D_i \cap D_j) \longrightarrow \cdots$$

with differentials

$$d^0((f_i)_i) = ((df_i)_i, (f_j - f_i)_{ij}, (-f_i|_{D_i})_i),$$
$$d^1((\omega_i)_i, (f_{ij})_{ij}, (g_i)_i)$$
$$= ((-\omega_j + \omega_i + df_{ij})_{ij}, (-f_{i_1 i_2} + f_{i_0 i_2} - f_{i_0 i_1})_{i_0 i_1 i_2}, (-f_{ij}|_{D_{ij}} - g_j + g_i)_{ij}).$$

Remark 3.21. The signs in the differentials depend on the sign conventions used for total complexes, shifts and cones. We are using the normalisations of [HMS17, Section 1.3]. However, any other choice of sign conventions will lead to isomorphic cohomology groups. We only have to ensure that $d^1 \circ d^0 = 0$.

3.2.2 The General Case

The definition of (relative) algebraic de Rham cohomology can be generalised to the singular case by different methods, all yielding the same cohomology groups. The first was Hartshorne's, who embedded a singular variety X into a smooth variety P and worked with the completion of the complex of sheaves of differential forms on P with respect to the vanishing ideal of X. In the context of Hodge theory, Deligne used smooth proper hypercovers: given X, he constructs a simplicial variety $X_\bullet$ over X with smooth components and such that the singular cohomology of X agrees with the singular cohomology of $X_\bullet$. (Our Proposition 3.12 is an instance of this fact.) He then defines algebraic de Rham cohomology of X as the de Rham cohomology of $X_\bullet$. In [HMS17], we use a variant of this approach. Algebraic de Rham cohomology is defined as sheaf cohomology of the complex Ω_h^* of h-differentials in the h-topology introduced by Voevodsky with the theory of triangulated motives in mind.

Instead of explaining the construction, we summarise the result:

Theorem 3.22 ([HMS17, Section 3.2]). *There is a sequence of functors*

$$H_{\mathrm{dR}}^n \colon (X, Y) \mapsto H_{\mathrm{dR}}^n(X, Y)$$

which attach a finite-dimensional k-vector space to every pair (X, Y) consisting of an algebraic variety and a closed subvariety Y and which extend the functors for smooth X and Y.

If $Z \subset Y \subset X$ are closed subvarieties, there are natural coboundary maps

$$\partial \colon H_{\mathrm{dR}}^n(Y, Z) \to H_{\mathrm{dR}}^{n+1}(X, Y)$$

fitting into a long exact sequence

$$\cdots \to H_{\mathrm{dR}}^n(X, Z) \to H_{\mathrm{dR}}^n(Y, Z) \xrightarrow{\partial} H_{\mathrm{dR}}^{n+1}(X, Y) \to \cdots.$$

Example 3.23. Let C be a curve with normalisation $\widetilde{C} \to C$ and $Y \subset C$ be the set of singular points with preimage $\widetilde{Y} \subset \widetilde{C}$. Then the shift by $[-1]$ of the cone of

$$R\widetilde{\Gamma}_{\mathrm{dR}}(\widetilde{C}) \oplus R\widetilde{\Gamma}_{\mathrm{dR}}(Y) \to R\widetilde{\Gamma}_{\mathrm{dR}}(\widetilde{Y})$$

computes algebraic de Rham cohomology of C. In particular, this leads to a short exact sequence

$$0 \to k^N \to H^1_{\mathrm{dR}}(C) \to H^1_{\mathrm{dR}}(\widetilde{C}) \to 0$$

with $N = |\widetilde{Y}| - |Y|$. This is the same result as in Example 3.13.

3.3 The Period Pairing

We now fix an embedding of k into $\mathbb{C}$, which allows us to define an analytification functor from k-varieties to complex spaces. Again we start under the simplifying assumption that X is smooth and affine. In this case there is the natural pairing

$$S_n^\infty(X^{\mathrm{an}}) \times \Omega^n(X) \to \mathbb{C}; \quad (\sigma, \omega) \mapsto \int_\sigma \omega.$$

By Stokes's Theorem the pairing induces a well-defined map on homology

$$H_n^{\mathrm{sing}}(X, \mathbb{Q}) \times H^n_{\mathrm{dR}}(X) \to \mathbb{C},$$

the *period pairing*.

Remark 3.24. For $n = 1$, which is the most important case for us, the pairing is even defined on all of $S_1(X)$. We get its value by analytic continuation.

For fixed n and (X, Y), singular cohomology is a $\mathbb{Q}$-vector space and de Rham cohomology is a k-vector space. After extension of scalars to $\mathbb{C}$, the pairing becomes perfect. In the smooth and proper case, this is a direct consequence of GAGA. In full generality, it was established by Deligne as part of the development of Hodge theory for non-proper and singular varieties. It even extends to pairs.

Theorem 3.25 ([HMS17, Chapter 5]). *There is a natural* period pairing

$$H_n^{\mathrm{sing}}(X, Y; \mathbb{Q}) \times H^n_{\mathrm{dR}}(X, Y) \to \mathbb{C}$$

inducing period isomorphisms

$$H^n_{\mathrm{sing}}(X, Y; \mathbb{Q}) \otimes_\mathbb{Q} \mathbb{C} \to H^n_{\mathrm{dR}}(X, Y) \otimes_k \mathbb{C}.$$

The period isomorphism is functorial for pairs of k-varieties and compatible with coboundary maps for triples $Z \subset Y \subset X$.

3.3.1 The Case of Smooth Affine Curves

For later use, we make the period pairing explicit in the first interesting case. Let C be a smooth affine curve, $Y \subset C$ a finite set of k-points. By definition, algebraic de Rham cohomology of (C, Y) is the cohomology of the complex

$$\left[\Omega^0(C) \to \Omega^1(C) \oplus \Omega^0(Y) \right].$$

Hence every class in $H^1_{\mathrm{dR}}(C, Y)$ is represented by a pair (ω, α) where ω is a 1-form and $\alpha \colon Y(k) \to k$ is a set-theoretic map. Conversely, every such pair defines a class in relative de Rham cohomology.

Singular homology of the pair is defined as homology of the complex

$$S^\infty_*(C)/S^\infty_*(Y).$$

Its 1-cycles are represented by $\mathbb{Z}$-linear combinations $\sigma = \sum n_i \gamma_i$ of smooth maps $\gamma_i \colon [0, 1] \to C^{\mathrm{an}}$ such that $\partial(\sum n_i \gamma_i) = \sum n_i \gamma_i(1) - \sum n_i \gamma_i(0)$ is in $S_0(Y)$. Up to homotopy such a cycle can be replaced by a formal $\mathbb{Z}$-linear combination of closed paths and paths with endpoints in $Y(k)$. The definition of the period pairing requires us to replace the singular complex of (C, Y) by the quasi-isomorphic cone,

$$\mathrm{cone}\left(S^\infty_*(Y) \to S^\infty(C)_* \right) = \left[S^\infty_0(C) \leftarrow S^\infty_1(C) \oplus S^\infty_0(Y) \leftarrow \cdots \right].$$

The homology class of σ is represented by the pair $(\sigma, -\partial\sigma)$. In these explicit terms, the period pairing is given by

$$\left((\omega, \alpha), (\sigma, \partial\sigma) \right) = \int_\sigma \omega - \alpha(\partial\sigma).$$

4

Commutative Algebraic Groups

4.1 The Building Blocks

We fix an algebraically closed field k of characteristic zero. The cases of relevance for us are $k = \overline{\mathbb{Q}}$ and $k = \mathbb{C}$. We denote by $\mathcal{G}$ the category of commutative connected algebraic groups over k. They are automatically smooth. This category is not abelian because the kernel of a morphism is not necessarily connected. However, the category of all commutative algebraic groups over k is abelian and so is the isogeny category of $\mathcal{G}$, where morphisms are tensored by $\mathbb{Q}$. For a careful review and analysis, see [Bri17].

Example 4.1. The additive group $\mathbb{G}_a = \mathrm{Spec}(k[T])$ and the multiplicative group $\mathbb{G}_m = \mathrm{Spec}(k[T, T^{-1}])$ are objects of $\mathcal{G}$.

A connected commutative algebraic group is called a *vector group* if it is isomorphic to a power of $\mathbb{G}_a$ and a *torus* if it is isomorphic to a power of $\mathbb{G}_m$.

Example 4.2. Every abelian variety over k is an object of $\mathcal{G}$.

Note that there are no non-trivial morphisms between vector groups, tori and abelian varieties.

Theorem 4.3 (Structure Theorem). *Let G be a connected commutative algebraic group. Then there is a canonical short exact sequence*

$$0 \to L \to G \to A \to 0$$

with an abelian variety A and a linear connected commutative algebraic group L. Moreover, there is a canonical split short exact sequence

$$0 \to V \to L \to T \to 0$$

with a torus T and a vector group V.

31

Proof The first sequence is the commutative case of [Bar55]; see also [Che60]. By [DG70, Chapter IV §3, Théorème 1.1] or [Ser88, Chapter III, Proposition 12] we have $L \cong V \times T$ with V unipotent and a torus T. By [Ser88, Chapter VII §2.7], all unipotent groups are powers of $\mathbb{G}_a$ in characteristic 0, hence V is a vector group. □

Corollary 4.4. *An object of $\mathcal{G}$ is simple if and only if it is isomorphic to $\mathbb{G}_a$, $\mathbb{G}_m$ or a simple abelian variety.*

A connected commutative algebraic group is called a *semi-abelian variety* if it is an extension of an abelian variety by a torus, or, equivalently, if its vector part is trivial. The Structure Theorem then implies that for any connected commutative algebraic group G there is also a canonical short exact sequence

$$0 \to V \to G \to G^{sa} \to 0$$

with a vector group V and a semi-abelian variety G^{sa}.

The aim of this chapter is on the one hand to give an alternative description of the category of semi-abelian varieties and, on the other hand, to explain the construction of the universal vector extension.

Definitions 4.5. 1. Let V be a vector group. We define $V^\vee = \mathrm{Hom}(V, k)$ and view it as a vector group.

2. Let A be an abelian variety. We define $A^\vee = \mathrm{Pic}^0(A)$, the *dual abelian variety*.

3. Let T be a torus. We define $X(T) = \mathrm{Hom}(T, \mathbb{G}_m)$, the *character group of T*.

4. Let Ξ be a free abelian group of finite rank. We define $\mathbb{G}_m(\Xi) = \mathrm{Hom}(\Xi, \mathbb{G}_m)$, the *dual torus of Ξ*.

Note that $A^\vee$ has a canonical structure of abelian variety by [Mum70, §13 Theorem]; see also [Mil08, p. 40]. All these functors are contravariant. Given a morphism of abelian varieties $\alpha\colon A \to A'$, by pull-back of line bundles we get an induced morphism $\alpha^\vee\colon A'^\vee \to A^\vee$. Given a morphism of tori $\tau\colon T \to T'$, by composition we get an induced morphism of abelian groups $\tau^*\colon X(T') \to X(T)$. Given a morphism of finitely generated abelian groups $\xi\colon \Xi \to \Xi'$, by composition we get an induced morphism of tori $\xi^*\colon \mathbb{G}_m(\Xi') \to \mathbb{G}_m(\Xi)$.

Proposition 4.6. *Let V be a vector group, A an abelian variety and T a torus. Then there are canonical isomorphisms*

$$V^{\vee,\vee} \cong V, \quad A^{\vee,\vee} \cong A, \quad \mathbb{G}_m(X(T)) \cong T.$$

Proof The case of V is linear algebra. In the case of T, by adjunction we get a canonical map

$$T \to \mathbb{G}_m(X(T)).$$

By naturality, it suffices to check that it is an isomorphism in the case $T = \mathbb{G}_m$. This case is again trivial. For the double dual of an abelian variety, see [Mum70, §13, Corollary, p. 132]; see also [Mil08, Theorem 8.9]. □

4.2 Group Extensions

In this section we give a short introduction to the functor Ext^1 in the abelian category of commutative algebraic groups defined over k. Most of the material can be found in Serre's book [Ser88].

Let A, B be objects of $\mathcal{G}$. To give an *extension of A by B* is the same as to give a triple (C, ι, π) with C a commutative algebraic group and $(\iota, \pi) \in \mathrm{Hom}(B, C) \times \mathrm{Hom}(B, A)$ such that

$$0 \longrightarrow B \overset{\iota}{\longrightarrow} C \overset{\pi}{\longrightarrow} A \longrightarrow 0 \qquad (4.1)$$

is exact. With abuse of notation, we often call the group C an *extension* of A by B. It is automatically connected.

A morphism of extensions is a triple $\alpha \colon A \to A', \beta \colon B \to B', \gamma \colon C \to C'$ making the diagram

$$
\begin{array}{ccccccccc}
0 & \longrightarrow & B & \overset{\iota}{\longrightarrow} & C & \overset{\pi}{\longrightarrow} & A & \longrightarrow & 0 \\
& & \downarrow{\scriptstyle \beta} & & \downarrow{\scriptstyle \gamma} & & \downarrow{\scriptstyle \alpha} & & \\
0 & \longrightarrow & B' & \overset{\iota'}{\longrightarrow} & C' & \overset{\pi'}{\longrightarrow} & A' & \longrightarrow & 0
\end{array}
\qquad (4.2)
$$

commutative. Note that γ is an isomorphism if and only if α and β are. In this case the extensions are isomorphic. We say that two extensions are *equivalent* if there is a homomorphism of extensions with $A = A'$, $B = B'$ and $\alpha = \mathrm{id}_A, \beta = \mathrm{id}_B$. The set of equivalence classes of extensions makes up a commutative group $\mathrm{Ext}^1(A, B)$, the group of *Yoneda-1-extensions*. By abuse of notation we often write C for its equivalence class $[C]$.

Remark 4.7. The morphisms α, β are uniquely determined by γ. If $\mathrm{Hom}(B, A') = 0$ (for example, because B is a linear algebraic group and A' an abelian variety), then the existence of α and β is automatic.

We now review the group structure on Ext^1 via the Baer sum. As for all abelian categories, the bi-functor Ext^1 which associates with a pair (A, B) the set $\mathrm{Ext}^1(A, B)$ is contravariant in the first and covariant in the second variable. The functoriality in the first variable is given by pull-back. Given a morphism $\alpha \colon A'' \to A$, introduce

$$C'' := C \times_A A''.$$

By construction,

$$0 \to B \to C'' \to A'' \to 0$$

is exact and we define

$$\alpha^*[C] = [C''] \in \mathrm{Ext}^1(A'', B).$$

The functoriality in the second variable is given by push-out. Given a morphism $\beta \colon B \to B'$, introduce

$$C' = C \times B'/B,$$

where B acts both on B' and on C. Let $B' \to C'$ be given by $(b, \beta(b))$. By construction,

$$0 \to B' \to C' \to A \to 0$$

is exact and we put

$$\beta_*[C] = [C'] \in \mathrm{Ext}^1(A, B').$$

The two transformations α^* and β_* commute in the sense that

$$\begin{array}{ccc}
\mathrm{Ext}^1(A, B) & \xrightarrow{\ \alpha^*\ } & \mathrm{Ext}^1(A'', B) \\
\downarrow{\scriptstyle \beta_*} & & \downarrow{\scriptstyle \beta^*} \\
\mathrm{Ext}^1(A, B') & \xrightarrow{\ \alpha^*\ } & \mathrm{Ext}(A'', B')
\end{array} \qquad (4.3)$$

commutes. This is a general property of Ext-groups in abelian categories.

If $[C_1]$ and $[C_2]$ are in $\mathrm{Ext}^1(A, B)$, then their *Baer sum* is

$$[C_1] + [C_2] = \Delta^* s_* ([C_1 \times C_2]),$$

which makes $\mathrm{Ext}^1(A, B)$ into a commutative group with neutral element $0 = [B \times A]$; here Δ is the diagonal map from A into $A \times A$ and $B \times B \xrightarrow{s} B$ the addition on B. We deduce that multiplication by an integer n can be defined inductively using addition. Equivalently take

$$n[B] = \Delta^* s_* ([B^n]),$$

where $\Delta = \Delta_n$ is the diagonal from C to C^n and $s = s_n$ is n-fold addition $A^n \to A$ on A.

The bi-functor Ext^1 is additive in both variables, which implies that

$$\mathrm{Ext}^1(A_1 \times A_2, B) = \mathrm{Ext}^1(A_1, B) \times \mathrm{Ext}^1(A_2, B)$$

and

$$\mathrm{Ext}^1(A, B_1 \times B_2) = \mathrm{Ext}^1(A, B_1) \times \mathrm{Ext}^1(A, B_2).$$

Hence it is of particular importance to understand Ext^1 for the simple building blocks.

Proposition 4.8. *Let A be an abelian variety and L, L' linear connected commutative algebraic groups. Then*

(1)
$$\text{Ext}^1(A, \mathbb{G}_a) = H^1(A, \mathcal{O}),$$
(2)
$$\text{Ext}^1(A, \mathbb{G}_m) = A^\vee(k) = \text{Pic}^0(A)(k) \subset \text{Pic}(A)(k) = H^1(A, \mathcal{O}^\times),$$
(3)
$$\text{Ext}^1(L, L') = 0.$$

Proof For the statements on abelian varieties, see [Ser88, Chapter VII §3, Theorems 7 and 6]. By Theorem 4.3, all linear connected commutative algebraic groups are split. In particular, there are no non-trivial extensions. □

Remark 4.9. The identification of $\text{Ext}^1(A, \mathbb{G}_m)$ with $A^\vee(k)$ is provided by the Poincaré bundle $\mathcal{P}$ on $A \times A^\vee$: see [Mum70, Chapter II §8, p. 78]. Given a point $x \in A^\vee(k)$, the pull-back $\mathcal{P}_x$ to A via (id, x) is a line bundle of degree 0. After removing the zero-section, it is a $\mathbb{G}_m$-bundle and, in fact, a semi-abelian variety in $\text{Ext}^1(A, \mathbb{G}_m)$.

Corollary 4.10. *Let G be a semi-abelian variety with abelian part A and V a vector group. Then the natural map*

$$\text{Ext}^1(A, V) \cong \text{Ext}^1(G, V)$$

is an isomorphism. In particular, $\text{Ext}^1(G, V)$ is finite dimensional.

Proof We start with the short exact sequence

$$0 \to T \to G \to A \to 0,$$

with T the torus part of G, and apply the long exact sequence for $\text{Hom}(-, V)$; see [Ser88, Chapter VII §1, no. 2]. This yields the exact sequence

$$\cdots \to \text{Hom}(T, V) \to \text{Ext}^1(A, V) \to \text{Ext}^1(G, V) \to \text{Ext}^1(T, V) \to \cdots.$$

The first and the last terms vanish.

Finite dimensionality holds because

$$\text{Ext}^1(A, V) \cong \text{Ext}^1(A, \mathbb{G}_a)^s \cong H^1(A, \mathcal{O})^s,$$

where $s = \dim(V)$. □

From this proposition, we get classifying maps: let A be an abelian variety and T a torus. Then we have the bilinear map

$$X(T) \times \mathrm{Ext}^1(A, T) \to \mathrm{Ext}^1(A, \mathbb{G}_m),$$
$$(\chi, [G]) \mapsto \chi_*[G].$$

On the other hand, let $[G] \in \mathrm{Ext}^1(A, T)$ and consider the exact sequence of Theorem 4.3 with $L = T$. Applying the long exact $\mathrm{Hom}(-, \mathbb{G}_m)$-sequence of [Ser88, Chapter VII §1, Proposition 2] we get a long exact sequence

$$\cdots \longrightarrow \mathrm{Hom}(T, \mathbb{G}_m) \xrightarrow{d_G} \mathrm{Ext}^1(A, \mathbb{G}_m) \xrightarrow{\pi^*} \mathrm{Ext}^1(G, \mathbb{G}_m) \longrightarrow \cdots,$$

where the connecting homomorphism is given by

$$d_G(\gamma) = \gamma_*(G).$$

The two descriptions are equivalent.

Corollary 4.11. *The induced map*

$$\mathrm{Ext}^1(A, T) \to \mathrm{Hom}(X(T), A^\vee)$$

is an isomorphism.

Proof As $T \cong \mathbb{G}_m^r$ and both sides are natural in T, it suffices to treat the case $T = \mathbb{G}_m$. In this case we have $X(\mathbb{G}_m) = \mathbb{Z}$ and $\mathrm{Hom}(X(\mathbb{G}_m), A^\vee) = A^\vee(k)$. The claim now follows from (2) in Proposition 4.8. $\qquad\qquad\square$

We refer to the image of G in $\mathrm{Hom}(X(T), A^\vee)$ as the *classifying map* of G.

Now let V be a vector group. Again we have a bilinear map

$$V^\vee \times \mathrm{Ext}^1(A, V) \to \mathrm{Ext}^1(A, \mathbb{G}_a),$$
$$(\lambda, [G]) \mapsto \lambda_*[G].$$

As in the torus case, there is an alternative description as

$$(\lambda, [G]) \mapsto d_G(\lambda)$$

with respect to the long exact $\mathrm{Hom}(-, \mathbb{G}_a)$-sequence attached to (4.1) with $L = V$.

Corollary 4.12. *The induced map*

$$\mathrm{Ext}^1(A, V) \to \mathrm{Hom}(V^\vee, H^1(A, \mathcal{O}))$$

is an isomorphism.

Proof As $V \cong \mathbb{G}_a^s$ and both sides are natural in V, it suffices to treat the case $V = \mathbb{G}_a$. In this case $\mathbb{G}_a^\vee = \mathbb{G}_a$ and $\mathrm{Hom}(\mathbb{G}_a, H^1(A, \mathcal{O})) = H^1(A, \mathcal{O})$. The claim follows from (1) in Proposition 4.8. □

The same considerations also apply to extensions of semi-abelian varieties by vector groups. We obtain:

Corollary 4.13. *Let G be semi-abelian and V a vector group. Then the natural map*

$$\mathrm{Ext}^1(G, V) \to \mathrm{Hom}(V^\vee, \mathrm{Ext}^1(G, \mathbb{G}_a))$$

is an isomorphism.

Proof By Corollary 4.10 we can replace G by its abelian part A on both sides. Then we are back in the situation of Corollary 4.12. □

4.3 Semi-abelian Varieties

As shown in Corollary 4.11, the datum of a semi-abelian variety G over k is equivalent to the datum of a homomorphism $X(T) \to A^\vee(k)$. This construction is functorial. A morphism of semi-abelian varieties $\alpha\colon G_1 \to G_2$ induces a commutative diagram

$$
\begin{array}{ccc}
X(T_1) & \xleftarrow{\ \alpha^\vee\ } & X(T_2) \\
{\scriptstyle [G_1]}\downarrow & & \downarrow{\scriptstyle [G_2]} \\
A_1^\vee(k) & \xleftarrow[\ \alpha^\vee\ ]{} & A_2^\vee(k).
\end{array}
$$

In conclusion, we have the following result.

Proposition 4.14. *The assignment $G \mapsto [X(T) \to A^\vee(k)]$ yields an equivalence between the category of semi-abelian varieties over k and the category with objects given by triples (X, A, ϕ) where X is a free abelian group of finite rank, A is an abelian variety and $\phi\colon X \to A^\vee(k)$ is a group homomorphism.*

Proof We verify that the functor is faithful. Let $f\colon G \to G'$ be a morphism of semi-abelian variety mapping to 0 under the functor. In particular, the induced morphisms on the torus part and the abelian part vanish. The composition $G \to G' \to A'$ vanishes because it factors over $0\colon A \to A'$. Hence f maps into T'. The restriction $f|_T$ vanishes, hence we get an induced map $\bar{f}\colon A \to T'$. There are no such maps, hence $\bar{f} = 0$.

The functor is also full: given a commutative diagram as above, we get back the morphism α as the composition

$$G_1 \to G_{\alpha^\vee \circ [G_2]} = G_2 \times_{A_2} A_1 \to G_2.$$

We are often going to make use of this equivalence without mentioning it explicitly. To verify the equality $G_{\alpha^\vee \circ [G_2]} = G_2 \times_{A_2} A_1$ we consider the cartesian diagram

$$
\begin{array}{ccc}
G_2 \times_{A_2} A_1 & \longrightarrow & G_2 \\
\downarrow & & \downarrow \\
A_1 & \longrightarrow & A_2,
\end{array}
$$

which corresponds to

$$
\begin{array}{ccc}
X(T_2) & = & X(T_2) \\
{\scriptstyle [G_2]}\downarrow & & \downarrow{\scriptstyle [G_2 \times_{A_2} A_1]} \\
A_2^\vee & \xrightarrow{\ \alpha^\vee\ } & A_1^\vee.
\end{array}
$$

This shows that $G_{\alpha^\vee \circ [G_2]} = G_2 \times_{A_2} A_1 = \alpha^* G_2$.

It remains to check that the functor is full on objects. Given (X, A, ϕ), we construct G as follows: let $e_1, \ldots, e_n$ be a basis of X. The elements $\phi(e_i) \in A^\vee(k)$ define elements of $\mathrm{Ext}^1(A, \mathbb{G}_m)$ and hence extensions

$$0 \to \mathbb{G}_m \to G_i \to A \to 0.$$

We put

$$G = G_1 \times_A \times \cdots \times_A G_n.$$

By construction, G maps to (X, A, ϕ) under our functor.   □

Remark 4.15. The map $X(T) \to A^\vee(k)$ is zero if and only if $G \cong A \times T$. Given two maps $s_1 \colon X(T_1) \to A^\vee(k)$ and $s_2 \colon X(T_2) \to A^\vee(k)$ corresponding to G_1 and G_2, their sum defines $s \colon X(T_1) \oplus X(T_2) \to A^\vee(k)$. It corresponds to the semi-abelian variety G obtained as the pull-back of $G_1 \times G_2 \to A \times A$ via the diagonal $A \to A \times A$. Its torus part is $T_1 \times T_2$. If $s_1 = 0$, then the composition $G \to G_1 \cong A \times T_1 \to T_1$ together with $G \to G_2$ induce an isomorphism $G \cong T_1 \times G_2$.

Definition 4.16. The category of semi-abelian varieties *up to isogeny* has the same objects as the category of semi-abelian varieties but with morphisms tensored by $\mathbb{Q}$.

Proposition 4.14 implies that the category of semi-abelian varieties up to isogeny is equivalent to the category with objects given by triples $(X_{\mathbb{Q}}, A, \phi)$ where $X_{\mathbb{Q}}$ is a finite-dimensional $\mathbb{Q}$-vector space, A denotes an abelian variety up to isogeny and ϕ a $\mathbb{Q}$-linear map $X_{\mathbb{Q}} \to A^{\vee}(k)_{\mathbb{Q}}$. We often write objects as $X \to A^{\vee}(k)_{\mathbb{Q}}$, where X is a free abelian group of finite rank.

Corollary 4.17. *Let G be semi-abelian variety, T a torus and $G \to T$ a surjective morphism of algebraic groups with kernel G'. Then $G \cong T \times G'$ up to isogeny.*

Proof Since tori are semi-simple there is also an injective homomorphism $T \to G$ with image in the torus part of G. By the universal property of the direct product, this leads to an isomorphism $G \cong T \times G'$. □

4.4 Universal Vector Extensions

As shown in Corollary 4.13, the datum of a vector extension of a semi-abelian variety G over k is equivalent to the datum of a linear map $V^{\vee} \to \mathrm{Ext}^1(G, \mathbb{G}_a)$ or dually $\mathrm{Ext}^1(G, \mathbb{G}_a)^{\vee} \to V$. As in the semi-abelian case, this construction is functorial.

Proposition 4.18. *The assignment $G \mapsto [\mathrm{Ext}^1(G, \mathbb{G}_a)^{\vee} \to V]$ yields an equivalence between the category of vector extensions of semi-abelian varieties over k and the category with objects given by triples (V, A, ϕ), where V is a finite-dimensional k-vector space, A is an abelian variety and $\phi: \mathrm{Ext}^1(G, \mathbb{G}_a)^{\vee} \to V$ is a k-linear map.*

Proof The argument is the same as in the semi-abelian case. □

The vector space $\mathrm{Ext}^1(G, \mathbb{G}_a)$ is itself finite dimensional by Corollary 4.10, hence there is a distinguished object in the category of vector extensions of A.

Definition 4.19. Let G be a semi-abelian variety. We call the vector extension

$$0 \to \mathrm{Ext}^1(G, \mathbb{G}_a)^{\vee} \to G^{\natural} \to G \to 0$$

corresponding to id: $\mathrm{Ext}^1(G, \mathbb{G}_a)^{\vee} \to \mathrm{Ext}^1(G, \mathbb{G}_a)^{\vee}$ the *universal vector extension of G.*

Proposition 4.20. *The universal vector extension of a semi-abelian variety G has the following universal property: given a vector extension*

$$0 \to V \to G' \to G \to 0,$$

there is a unique morphism $G^\natural \to G'$ compatible with the projection to G.

Proof Under the equivalence of Proposition 4.18 the vector extension G' corresponds to the triple $(V, G, \phi \colon \mathrm{Ext}^1(G, \mathbb{G}_a)^\vee \to V)$ and $G^\natural$ corresponds to $(\mathrm{Ext}^1(G, \mathbb{G}_a), A, \mathrm{id})$. A morphism $G' \to G^\natural$ corresponds to a linear map $\mathrm{Ext}^1(G, \mathbb{G}_a)^\vee \to V$ compatible with the structure maps. The only such map is ϕ. □

Remark 4.21. Let A be the abelian part of G. By the computation of $\mathrm{Ext}^1(A, \mathbb{G}_a)$ in Proposition 4.8, we have

$$0 \to H^1(A, \mathcal{O})^\vee \to A^\natural \to A \to 0.$$

Moreover, the isomorphism $\mathrm{Ext}^1(A, \mathbb{G}_a) \cong \mathrm{Ext}^1(G, \mathbb{G}_a)$ of Corollary 4.10 implies that $G^\natural$ is explicitly constructed as

$$G^\natural = A^\natural \times_A G.$$

If V is a vector group, $G \in \mathcal{G}$, both $\mathrm{Hom}(G, V)$ and $\mathrm{Ext}^1(G, V)$ are k-vector spaces and do not change when we replace the category $\mathcal{G}$ with its isogeny category $\mathcal{G}_\mathbb{Q}$. This remark implies the following result.

Corollary 4.22. *The universal vector extension $G^\natural$ of a semi-abelian variety G also satisfies the universal property of a vector extension in the isogeny category $\mathcal{G}_\mathbb{Q}$.*

4.5 Generalised Jacobians

Let k be an algebraically closed field of characteristic zero. Let Y be a smooth connected algebraic curve over k with a chosen base point y_0.

The following theorem is a special case of the theory of generalised Jacobians introduced by Rosenlicht. We follow the presentation of Serre; see [Ser88, Chapter V]. We recall briefly the deduction.

Theorem 4.23 (Rosenlicht: see Serre [Ser88, Chapter V]). *There is a semi-abelian variety $J(Y)$ over k and a morphism*

$$Y \to J(Y)$$

depending only on y_0 such that $H_1^{\mathrm{sing}}(Y, \mathbb{Z}) \to H_1^{\mathrm{sing}}(J(Y), \mathbb{Z})$ is an isomorphism.

4.5.1 Construction of $J(Y)$

Let $\bar{Y}$ be the smooth compactification closure of Y and $S = \bar{Y} \setminus Y$ the set of points in the complement of Y. We define the divisor $\mathfrak{m}$ as $\sum_{P \in S} P$. In the terminology of [Ser88] this is a (special case of a) *modulus*. The case $\mathfrak{m} = 0$ (i.e. $S = \varnothing$) is allowed.

A rational function φ on Y is congruent to 1 mod $\mathfrak{m}$ if $v_P(1 - \varphi) \geq 1$ for all $P \in S$ where v_P denotes the valuation at P. We write

- $C_{\mathfrak{m}}$ for the group of classes of divisors on $\bar{Y}$ which are prime to S modulo those which can be written as (φ) for some rational function $\varphi \equiv 1 \mod \mathfrak{m}$;
- $J_{\mathfrak{m}} = C_{\mathfrak{m}}^0$ for the subgroup of classes which have degree 0;
- $J = C^0$ for the usual group of divisor classes of degree 0.

There is a surjective homomorphism

$$\pi \colon J_{\mathfrak{m}} \to J$$

with kernel $L_{\mathfrak{m}}$ consisting of those classes in $J_{\mathfrak{m}}$ which are invertible at each $P \in S$. Moreover, let

$$\theta \colon Y \to J_{\mathfrak{m}}$$

be the map assigning to a point $y \in Y$ the class of the divisor $y - y_0$.

Alternatively, the group $J_{\mathfrak{m}}$ can be described as the group of isomorphism classes of pairs (L, ι) where L is a line bundle of degree 0 on $\bar{Y}$ and ι is a trivialisation of L on S. The image of a divisor D of degree 0 on $\bar{Y}$ is the line bundle $\mathcal{O}(D)$ together with the canonical trivialisation, which exists because $\mathcal{O}(D)|_U = \mathcal{O}|_U$ outside the support of the divisor, in particular on S. In the case $S = \varnothing$, this identification is the familiar isomorphism $J \cong \mathrm{Pic}^0(\bar{Y})$.

By [Ser88, Chapter V, Proposition 2], the group $J_{\mathfrak{m}}$ is an algebraic group and by Serre's Proposition 4 the map θ is a morphism of algebraic varieties. By [Ser88, Chapter V, Theorem 2], the pair has a universal property for morphisms into commutative algebraic groups mapping y_0 to 0: given a rational map $f \colon \bar{Y} \to G$ to a commutative algebraic group G admitting $\mathfrak{m}$ for a modulus; see [Ser88, Chapter I, Theorem 1], there exists a unique algebraic homomorphism $F \colon J_{\mathfrak{m}} \to G$ such that

$$f = F \circ \theta + f(y_0).$$

The structure of $J_{\mathfrak{m}}$ is explained in [Ser88, Chapter V, Section 13]. In the case $\mathfrak{m} = 0$, we get back the usual Jacobian of $\bar{C}$. This is an abelian variety. In our special case, the kernel $L_{\mathfrak{m}}$ is isomorphic to $\mathbb{G}_m^r$ where

$$r = \begin{cases} 0 & \text{for } \mathfrak{m} = 0, \\ \deg \mathfrak{m} - 1 & \text{for } \mathfrak{m} \neq 0. \end{cases}$$

We put $J(Y) := J_{\mathfrak{m}}$ and obtain up to isomorphism the short exact sequence

$$1 \to \mathbb{G}_m^r \to J(Y) \to J(\bar{Y}) \to 0$$

of commutative algebraic groups, making $J(Y)$ semi-abelian.

The classifying map of $J(Y)$ maps a lattice of rank r to $J(\bar{Y})^\vee \cong \operatorname{Pic}^0(\bar{Y}) \cong J(\bar{Y})$.

Lemma 4.24 (Serre [Ser60, Section 1]). *The classifying map of $J(Y)$ is given by the map*

$$\mathbb{Z}[S]^0 \to J(\bar{Y}) \cong J(\bar{Y})^\vee$$

induced by θ where $\mathbb{Z}[S]^0$ is the group of divisors of degree 0 supported on the S.

4.5.2 Generalised Jacobian over $\mathbb{C}$

The structure of $J_{\mathfrak{m}}$ over $\mathbb{C}$ is explained in [Ser88, Chapter V §19]. Serre shows that

$$J_{\mathfrak{m}}(\mathbb{C}) \cong H^0(\bar{Y}, \Omega^1(-\mathfrak{m}))^\vee / H_1^{\text{sing}}(Y, \mathbb{Z}).$$

This implies that the map induced by $Y \to J(Y)$ induces an isomorphism

$$H_1^{\text{sing}}(Y, \mathbb{Z}) \to H_1^{\text{sing}}(J(Y)(\mathbb{C}), \mathbb{Z}).$$

Remark 4.25. Actually, this isomorphism is shown in [Ser88] on the way to establishing the formula for $J_{\mathfrak{m}}(C)$.

5

Lie Groups

We review the construction and properties of the exponential map, establishing notation and normalisations for later.

5.1 The Lie Algebra

Let G^{an} be a connected commutative complex Lie group. We denote by $\mathfrak{g}_{\mathbb{C}}$ or $\mathrm{Lie}(G^{\mathrm{an}})$ the Lie algebra of invariant vector fields on G^{an} and by $\mathfrak{g}_{\mathbb{C}}^{\vee}$ or $\mathrm{coLie}(G^{\mathrm{an}})$ the dual space of invariant differential forms. Note that $\mathfrak{g}_{\mathbb{C}}$ is abelian, i.e. the Lie bracket is trivial and does not play a role in what follows.

If V is a $\mathbb{C}$-vector space, we can view it as a complex commutative Lie group V^{an}. In this case $\mathrm{Lie}(V^{\mathrm{an}}) = V$.

Example 5.1. For $\mathbb{G}_a = \mathrm{Spec}(\mathbb{Z}[t])$ we have $\mathbb{G}_a^{\mathrm{an}} = \mathbb{C}$. It has a canonical coordinate with the property $t(1) = 1$. Then its Lie algebra $\mathfrak{g}_a^{\mathrm{an}}$ is generated by d/dt and its dual by dt. The canonical identification of $\mathfrak{g}_a^{\mathrm{an}}$ with $\mathbb{G}_a^{\mathrm{an}}$ maps d/dt to 1.

Morphisms in the category of connected commutative complex Lie groups are called *analytic homomorphisms*. The assignments Lie and coLie are functors. Given an analytic homomorphism $\varphi\colon G^{\mathrm{an}} \to H^{\mathrm{an}}$ of connected commutative complex Lie groups, by push-forward of vector fields and pull-back of differential forms we get $\mathbb{C}$-linear maps

$$\varphi_* = d\varphi\colon \mathfrak{g}_{\mathbb{C}} \to \mathfrak{h}_{\mathbb{C}}, \quad \varphi^* = \delta\phi\colon \mathfrak{h}_{\mathbb{C}}^{\vee} \to \mathfrak{g}_{\mathbb{C}}^{\vee}.$$

The linear maps are adjoint with respect to the natural pairings between a space and its dual space and this means that $(\varphi_*)^{\vee} = \varphi^*$. If $(\ ,\ )$ is the pairing which defines duality, then

$$\big(\varphi^*(\omega), X\big) = \big(\omega, \varphi_* X\big)$$

for every invariant differential form ω on H and invariant vector field X on G.

Of particular interest is the case of analytic homomorphisms $\varphi \colon \mathbb{C} \to G^{\text{an}}$ and $X = d/dt$.

5.2 The Exponential Map

For any given vector field $X \in \mathfrak{g}_{\mathbb{C}}$ there exists a unique analytic homomorphism $\varphi_X \colon \mathbb{G}_a^{\text{an}} \to G^{\text{an}}$ such that its tangent map $d\varphi_X$ satisfies

$$(d\varphi_X)\left(\frac{d}{dt}\right) = X,$$

as can be found in [War83]. It amounts to solving an ordinary linear differential equation. This is used to construct the exponential map of the Lie group G^{an} in the following way. We choose $X \in \mathfrak{g}_{\mathbb{C}}$ and put

$$\exp_G(X) := \varphi_X(1).$$

This defines an analytic homomorphism $\exp_G \colon \mathfrak{g}_{\mathbb{C}} \to G^{\text{an}}$. In other words, $\exp_G$ is uniquely characterised by functoriality with respect to analytic group homomorphisms and $d\exp_G = \text{id}$. This leads to an exact sequence

$$0 \longrightarrow \Lambda \longrightarrow \mathfrak{g}_{\mathbb{C}} \xrightarrow{\exp_G} G^{\text{an}} \longrightarrow 0. \qquad (5.1)$$

Proposition 5.2. *Let G^{an} be a connected commutative complex Lie group. Then $\exp_G \colon \mathfrak{g}_{\mathbb{C}} \to G^{\text{an}}$ is the universal cover.*

Proof The map $\exp_G$ is unramified because $d\exp = \text{id}$ is an isomorphism. It is a cover by the sequence (5.1). This makes it the universal cover. □

Remark 5.3. This also means that Λ is the fundamental group of G^{an}. We will deduce an explicit identification from the point of view of paths below.

An analytic homomorphism $\varphi \colon G^{\text{an}} \to H^{\text{an}}$ induces a commutative diagram

$$
\begin{array}{ccccccccc}
0 & \longrightarrow & \Lambda_G & \xrightarrow{\iota_G} & \mathfrak{g}_{\mathbb{C}} & \xrightarrow{\exp_G} & G^{\text{an}} & \longrightarrow & 0 \\
& & \downarrow{\scriptstyle \nu} & & \downarrow{\scriptstyle d\varphi} & & \downarrow{\scriptstyle \varphi} & & \\
0 & \longrightarrow & \Lambda_H & \xrightarrow{\iota_H} & \mathfrak{h}_{\mathbb{C}} & \xrightarrow{\exp} & H^{\text{an}} & \longrightarrow & 0.
\end{array}
\qquad (5.2)
$$

The converse does not hold necessarily: a Lie algebra homomorphism $\theta \colon \mathfrak{g}_{\mathbb{C}} \to \mathfrak{h}_{\mathbb{C}}$ does not necessarily descend to an analytic homomorphism from G^{an} to H^{an}. However, this does hold if G^{an} is simply connected:

Lemma 5.4. *Let G^{an} be simply connected, $\theta: \mathfrak{g}_{\mathbb{C}} \to \mathfrak{h}_{\mathbb{C}}$ a linear map. Then there exists an analytic homomorphism $\Theta: G^{an} \to H^{an}$ such that $d\Theta = \theta$.*

Proof As G^{an} is simply connected, it agrees with its universal cover. Hence $\exp_G$ is an isomorphism of connected commutative complex Lie groups. We define

$$\Theta = \exp_H \circ \theta \circ \exp_G^{-1}. \qquad \square$$

The group G^{an} is simply connected if and only if G^{an} is a vector group $V \cong \mathbb{G}_a^n$.

5.3 Integration over Paths

Another pairing is obtained by integration. Let $\gamma: [0,1] \to G^{an}$ be any path in G^{an}. This path defines an element $I(\gamma)$ in $\mathfrak{g}_{\mathbb{C}}$ by putting

$$I(\gamma)(\omega) := \int_{\gamma} \omega \quad \text{for all } \omega \in \mathfrak{g}_{\mathbb{C}}^{\vee}.$$

By Stokes's Theorem we see that $I(\gamma)$ depends only on the homotopy class of γ. The pairing is non-degenerate so $I(\gamma) = 0$ implies that γ is closed and homotopically equivalent relative $\{0,1\}$ to a constant path.

Example 5.5. In the case that $G = \mathbb{G}_a$ and $\epsilon: [0,1] \to \mathbb{G}_a^{an}$ is the path going from $\epsilon(0) = 0$ straight to $\epsilon(1) = 1$, we have

$$I(\epsilon) = \left(-, \frac{d}{dt}\right).$$

In fact every invariant differential form on $\mathbb{G}_a$ is a constant multiple of dt and everything reduces to the calculation

$$\int_0^1 dt = 1 = \left(dt, \frac{d}{dt}\right).$$

Let $\varphi: G^{an} \to H^{an}$ be an analytic homomorphism. Then we have for all invariant differential forms $\omega \in \mathfrak{h}_{\mathbb{C}}$ and paths γ on G^{an},

$$I(\varphi_* \gamma)(\omega) = I(\gamma)(\varphi^* \omega)$$

by the transformation rule.

Now fix an element X in $\mathfrak{g}_{\mathbb{C}}$ and let φ_X be the analytic homomorphism from $\mathbb{G}_a^{an}$ to G^{an} determined by X. Also let

$$\gamma_X: [0,1] \to G^{an}$$

be the path obtained by restricting the analytic homomorphism φ_X to the interval $[0,1]$. Note that by definition $\gamma_X = \varphi_X \circ \epsilon = \varphi_{X,*}\epsilon$.

We have thus defined maps I and $X \mapsto \gamma_X$, assigning tangent vectors to paths and conversely.

Lemma 5.6. *We have* $I(\gamma_X) = X$.

Proof Let ω be in $\mathfrak{g}_{\mathbb{C}}^{\vee}$. Since γ_X is a restriction of the analytic homomorphism φ_X we see that $\gamma_X^*\omega$ is an invariant differential form in $\mathfrak{g}_{a,\mathbb{C}}$. Then

$$I(\gamma_X)(\omega) = I(\varphi_{X,*}\epsilon)(\omega) = I(\epsilon)(\varphi_X^*\omega) = \left(\varphi_X^*\omega, \frac{d}{dt}\right) = \left(\omega, \varphi_X^*\frac{d}{dt}\right) = (\omega, X)$$

by the transformation formula for integrals, together with $I(\epsilon) = (-, d/dt)$. This means that $I(\gamma_X) = X$. $\qquad\square$

In particular, we may start with the element $X = I(\gamma)$. Then

$$I(\gamma) = X = I(\gamma_X)) = I(\gamma_{I(\gamma)}),$$

whence γ is homotopic to $\gamma_{I(\gamma)}$. This gives $\gamma(1) = \gamma_{I(\gamma)}(1) = \exp_G(I(\gamma))$ and leads to the following

Corollary 5.7. *Let P be a point in G^{an} and γ a path from 0 to P. Then we have*

$$\exp_G(I(\gamma)) = P.$$

The lemma shows that integration is inverse to exponentiation, as it should be. But this is precisely the definition of a logarithm, and we may write

$$\log_G(P) := I(\gamma).$$

Note that $\log_G$ is multi-valued. The map $\gamma \mapsto I(\gamma)$ from the path space $\mathcal{L}_G(0)$ of G, with the unit element of the group as base point, taken modulo homotopy into the Lie algebra $\mathfrak{g}_{\mathbb{C}}$ identifies $\mathfrak{g}_{\mathbb{C}}$ with the universal covering space of G^{an}.

We now restrict the discussion to closed paths. The maps

$$\Lambda \underset{I}{\overset{X \mapsto \gamma_X}{\rightleftarrows}} \pi_1(G^{\mathrm{an}}, 0) \tag{5.3}$$

are inverse to each other; in particular, $\Lambda \cong \pi_1(G, 0)$ and the fundamental group is abelian.

Let $\sigma = \sum_{i=1}^{n} a_i\gamma_i$ be a chain with $a_i \in \mathbb{Z}$, $\gamma_i \colon [0,1] \to G^{\mathrm{an}}$ continuous. We extend the definition of I and put

$$I(\sigma) = \sum_{i=1}^{n} a_i I(\gamma_i) \in \mathfrak{g}_{\mathbb{C}}.$$

If γ is closed, but $\gamma(0) \neq 0$, then $p(I(\gamma))$ is still homologous to γ. Hence the maps

$$\Lambda \underset{I}{\overset{X \mapsto \gamma_X}{\rightleftarrows}} H_1^{\mathrm{sing}}(G^{\mathrm{an}}, \mathbb{Z}) \tag{5.4}$$

are inverse to each other. The two identifications are compatible with the Hurewitz map $\pi_1(G^{\mathrm{an}}, 0) \to H_1^{\mathrm{sing}}(G^{\mathrm{an}}, \mathbb{Z})$, which is an isomorphism in this case.

6

The Analytic Subgroup Theorem

In this chapter, we give a new formulation of the Analytic Subgroup Theorem. We then explore its consequences for the comparison of analytic and algebraic homomorphisms.

6.1 The Statement

Let G be a commutative and connected algebraic group defined over $\overline{\mathbb{Q}}$ and $\mathfrak{g}$ its Lie algebra. The associated complex manifold G^{an} is a complex Lie group and its Lie algebra is $\mathfrak{g}_{\mathbb{C}} = \mathfrak{g} \otimes_{\overline{\mathbb{Q}}} \mathbb{C}$. The exponential map

$$\exp_G : \mathfrak{g}_{\mathbb{C}} \to G^{\mathrm{an}}$$

from the Lie algebra into G^{an} defines an analytic homomorphism. If $\mathfrak{b} \subseteq \mathfrak{g}$ is a subalgebra and $\mathfrak{b}_{\mathbb{C}} = \mathfrak{b}_{\overline{\mathbb{Q}}} \otimes \mathbb{C}$ we denote by B the analytic subgroup $\exp_G(\mathfrak{b}_{\mathbb{C}})$. An obvious question one can ask is whether $B(\overline{\mathbb{Q}}) := B \cap G(\overline{\mathbb{Q}})$ can contain an algebraic point different from 0, the neutral element. The answer is given by the Analytic Subgroup Theorem.

Theorem 6.1 (Wüstholz [Wüs87, Wüs89]). *The group of algebraic points* $B(\overline{\mathbb{Q}})$ *is non-trivial if and only if there is a connected algebraic subgroup* $H \subseteq G$ *with Lie algebra* $\mathfrak{h}$ *such that* $\{0\} \neq \mathfrak{h} \subseteq \mathfrak{b}$.

We conclude that the only source for algebraic points is the obvious one. Note that $B(\overline{\mathbb{Q}}) \neq \{0\}$ implies that $\mathfrak{b} \neq \{0\}$.

There is a refined version of the theorem. To state it, let G be a connected commutative algebraic group over $\overline{\mathbb{Q}}$ with Lie subalgebra $\mathfrak{g}$ and let $\langle \, , \, \rangle$ be the duality pairing between $\mathfrak{g}^{\vee}$ and $\mathfrak{g}$. For u in $\mathfrak{g}_{\mathbb{C}}$ with $\exp_G(u) \in G(\overline{\mathbb{Q}})$ we denote by $\mathrm{Ann}(u)$ the largest subspace of $\mathfrak{g}^{\vee}$ (sic) such that $\langle \mathrm{Ann}(u), u \rangle = 0$.

Theorem 6.2. *Assume that* $\exp_G(u) \in G(\overline{\mathbb{Q}})$. *Then there exists an exact sequence*

$$0 \to H \to G \xrightarrow{\pi} G/H \to 0$$

of connected commutative algebraic groups defined over $\overline{\mathbb{Q}}$ *such that* $\mathrm{Ann}(u) = \pi^*(\mathfrak{g}/\mathfrak{h})^{\vee}$ *and* $u \in \mathfrak{h}_{\mathbb{C}}$, *where* $\mathfrak{h}$ *is the Lie algebra of H. The sequence is uniquely determined by these properties.*

Proof We write $P = \exp_G(u)$. If $u = 0$, the theorem holds with $H = 0$. If $u \neq 0$, but $P = \exp_G(u) = 0$, we may replace u by u/n for a big enough $n \in \mathbb{N}$. We then have $\exp_G(u/n) \neq 0$ because the kernel of $\exp_G$ is discrete. Moreover, the image point is a torsion point of G, hence in $G(\overline{\mathbb{Q}})$. Without loss of generality, we may assume that $P \neq 0$.

Let $\langle \ , \ \rangle \colon \mathfrak{g}^{\vee} \times \mathfrak{g} \to \overline{\mathbb{Q}}$ be the natural duality pairing and for any subalgebra $\mathfrak{a} \subset \mathfrak{g}$ denote the left kernel by

$$\mathfrak{a}^{\perp} = \{\lambda \in \mathfrak{g}^{\vee}; \langle \lambda, \mathfrak{a} \rangle = 0\}.$$

The right kernel is defined similarly. We put $\mathfrak{h} := \mathrm{Ann}(u)^{\perp} \subseteq \mathfrak{g}$. Then $\mathfrak{h}_{\mathbb{C}}$ contains u and Theorem 6.1 gives an algebraic subgroup $H \subset G$ with Lie algebra $\mathfrak{h}$. We may assume that $u \in \mathfrak{h}_{\mathbb{C}}$, otherwise we apply our arguments to G/H. It has smaller dimension, so the process must stop after finitely many steps. Taking the left kernels gives $\mathfrak{h}^{\perp} \subset \mathfrak{h}^{\perp} \subset \mathrm{Ann}(u)$ and then $\mathfrak{h}^{\perp} = \mathrm{Ann}(u) = \mathfrak{h}^{\perp}$ by the maximality property of $\mathrm{Ann}(u)$. We get an exact sequence of Lie algebras

$$0 \to \mathfrak{h} \to \mathfrak{g} \xrightarrow{\pi_*} \mathfrak{g}/\mathfrak{h} \to 0,$$

which corresponds to an exact sequence

$$0 \to H \to G \xrightarrow{\pi} G/H \to 0$$

of algebraic groups and by duality to the exact sequence

$$0 \to (\mathfrak{g}/\mathfrak{h})^{\vee} \xrightarrow{\pi^*} \mathfrak{g}^{\vee} \to \mathfrak{h}^{\vee} \to 0.$$

We have $\mathfrak{h}^{\perp} = \pi^*(\mathfrak{g}/\mathfrak{h})^{\vee}$ and $(\mathfrak{g}/\mathfrak{h})^{\perp} = \mathfrak{h}^{\vee}$, which we prove as follows: we have $\lambda \in \mathfrak{h}^{\perp}$ if and only if the restriction of λ to $\mathfrak{h}$ is zero. This implies that λ descends to $\mathfrak{g}/\mathfrak{h}$ and that there is an element $\mu \in (\mathfrak{g}/\mathfrak{h})^{\vee}$ with $\lambda = \pi^*\mu$. This leads to $\mathfrak{h}^{\perp} \subseteq \pi^*(\mathfrak{g}/\mathfrak{h})^{\vee}$. Conversely

$$\langle \pi^*(\mathfrak{g}/\mathfrak{h})^{\vee}, \mathfrak{h} \rangle = \langle (\mathfrak{g}/\mathfrak{h})^{\vee}, \pi_*\mathfrak{h} \rangle = 0$$

since $\pi_*\mathfrak{h} = 0$, and we conclude that $\mathrm{Ann}(u) = \mathfrak{h}^{\perp} = \pi^*(\mathfrak{g}/\mathfrak{h})^{\vee}$ as stated.

Suppose that there is another short exact sequence

$$0 \to H' \to G \xrightarrow{\pi'} G/H' \to 0$$

with the same properties. In particular $\pi'^*(\mathfrak{g}/\mathfrak{h})^\vee = \pi^*(\mathfrak{g}/\mathfrak{h}')^\vee$ as subobjects of $\mathfrak{g}^\vee$. This implies that $\mathfrak{h} = \mathfrak{h}'$ as subspaces of $\mathfrak{g}$. As H and H' are connected, this also implies $H = H'$ as subgroups of G. □

6.2 Analytic vs Algebraic Homomorphisms

The Subgroup Theorem also has a consequence for the category of groups itself. A connected commutative algebraic group over $\overline{\mathbb{Q}}$ gives rise to a complex Lie group. We recall that morphisms in the category of complex Lie groups are called *analytic homomorphisms*.

Theorem 6.3. *Let G, G' be connected commutative algebraic groups defined over $\overline{\mathbb{Q}}$ with Lie algebras $\mathfrak{g}$ and $\mathfrak{g}'$ and let $\phi: G^{an} \to G'^{an}$ be an analytic group homomorphism such that $\mathfrak{g}_{\mathbb{C}} \to \mathfrak{g}'_{\mathbb{C}}$ maps $\mathfrak{g}$ to $\mathfrak{g}'$.*

Then there exists a vector group V_{tr}, a connected commutative algebraic group G_{alg} and a decomposition

$$G \cong V_{tr} \times G_{alg}$$

such that $\phi|_{G_{alg}^{an}}$ is algebraic over $\overline{\mathbb{Q}}$ and $\phi|_{V_{tr}^{an}}$ is purely transcendental, i.e. $\phi(V_{tr}(\overline{\mathbb{Q}}))$ does not contain any non-zero algebraic values.

Remarks 6.4. 1. An earlier version made the same claim but without the V_{tr}-factor. We thank the referee for pointing out the mistake in the argument. Indeed, the statement would be false, as the example $\exp: \mathbb{C} \to \mathbb{C}^*$ shows. The theorem states that all counterexamples are of a similar nature; see Corollary 6.9 for a complete classification.

2. The assumption on the induced map on Lie algebras is necessary as the following example shows: let $G_1 = \mathbb{G}_m$ and let $G_2 = E$ be an elliptic curve over $\overline{\mathbb{Q}}$. Let

$$\mathbb{C} \xrightarrow{z \mapsto \exp(2\pi i z)} \mathbb{C}^* = \mathbb{G}_m^{an}$$

be the standard uniformisation. For E^{an} we use the explicit uniformisation

$$\exp_E: \mathbb{C} \to E^{an}$$

with kernel $\Lambda = \mathbb{Z}\omega_1 + \mathbb{Z}\omega_2$ of Section 18.1. In these coordinates the $\overline{\mathbb{Q}}$-coLie algebras of $\mathbb{G}_m$ and E are generated by dz/z and dz respectively.

We get a well-defined analytic homomorphism

$$\phi: \mathbb{G}_m^{an} = \mathbb{C}/2\pi i \mathbb{Z} \to E^{an} = \mathbb{C}/\Lambda$$

by mapping $z \mapsto (\omega_1/2\pi i)z$. It is not algebraic. Note that this does not contradict Theorem 6.3 because ϕ does not map $\mathfrak{g}_1$ to $\mathfrak{g}_2$ since $\omega_1/2\pi i$ is not in $\overline{\mathbb{Q}}$, as we shall see later.

The proof of this theorem will take the rest of this chapter.

Lemma 6.5. *Suppose that the set of torsion points G_{tor} is dense in G. Under the assumptions of Theorem 6.3, the morphism ϕ is algebraic.*

Proof Let $B \subset G^{an} \times G'^{an}$ be the graph of ϕ. It is connected because it is isomorphic to G^{an} via the first projection. By assumption its Lie algebra is defined over $\overline{\mathbb{Q}}$. Let $g \in G(\overline{\mathbb{Q}})$ be an N-torsion point. Then $\phi(g) \in G'(\mathbb{C})$ is also an N-torsion point, hence in $G'(\overline{\mathbb{Q}})$. This implies that

$$T := \{(g, \phi(g)) | g \in G_{tor}\} \subset B(\overline{\mathbb{Q}}).$$

By the Analytic Subgroup Theorem there is an algebraic subgroup $H \subset G \times G'$ defined over $\overline{\mathbb{Q}}$ such that $H^{an} \subset B$ and containing all of T. The projection $B \hookrightarrow G^{an} \times G'^{an} \to G^{an}$ is an isomorphism, hence its restriction to H is a closed immersion. The image contains the set G_{tor}. It is Zariski dense, hence the inclusion is surjective. In other words, $H^{an} = B$. The group $H \subset G \times G'$ is the graph of the morphism we wanted to find. □

Lemma 6.6. *The theorem holds if $G = V$ is a vector group.*

Proof Let $\Sigma = \phi^{-1}(G'(\overline{\mathbb{Q}})) \cap V(\overline{\mathbb{Q}})$. We denote by $V_\Sigma \subset V$ the smallest algebraic subgroup containing Σ. We choose a direct complement V_{tr} of V_Σ in V. By construction, $\phi|_{V_{tr}^{an}}$ is purely transcendental. Indeed, any $\sigma \in V_{tr}(\overline{\mathbb{Q}})$ with $\phi(\sigma) \in G'(\overline{\mathbb{Q}})$ is already in Σ and hence in $V_\Sigma(\overline{\mathbb{Q}})$.

It remains to show that $\phi|_{V_\Sigma^{an}}$ is algebraic. As in the previous lemma, we consider its graph B in $V_\Sigma^{an} \times G'^{an}$. Its Lie algebra is defined over $\overline{\mathbb{Q}}$ and it contains the set

$$T := \{(g, \phi(g)) | g \in \Sigma\} \subset B(\overline{\mathbb{Q}}).$$

By the Analytic Subgroup Theorem there is an algebraic subgroup $H \subset V_\Sigma \times G'$ defined over $\overline{\mathbb{Q}}$ such that $H^{an} \subset B$ and containing all of T. The projection $B \hookrightarrow V_\Sigma^{an} \times G'^{an} \to V_\Sigma^{an}$ is an isomorphism, hence its restriction to H is a closed immersion. The image is an algebraic subgroup containing the set Σ, hence equal to V_Σ. In other words, again $H^{an} = B$. The group $H \subset G \times G'$ is the graph of the morphism we wanted to find. □

Lemma 6.7. *Let $G_1 \to G_2$ be a vector extension. Then $(G_1)_{tor} = (G_2)_{tor}$.*

Proof It suffices to check the statement over the complex numbers and in the analytification. We have

$$G_1^{\text{an}} \cong \mathbb{C}^{n_1} / H_1^{\text{sing}}(G_1^{\text{an}}, \mathbb{Z}) \to G_2^{\text{an}} \cong \mathbb{C}^{n_2} / H_1^{\text{sing}}(G_2^{\text{an}}, \mathbb{Z}).$$

By homotopy invariance, $H_1^{\text{sing}}(G_1^{\text{an}}, \mathbb{Z}) \cong H_1^{\text{sing}}(G_2^{\text{an}}, \mathbb{Z})$. The torsion is computed as $G_{i, \text{tor}} \cong H_1^{\text{sing}}(G_i^{\text{an}}, \mathbb{Z}) \otimes \mathbb{Q}/\mathbb{Z}$, hence the torsions of G_1 and G_2 are isomorphic. □

Let V be the vector part of G, i.e. we have a short exact sequence

$$0 \to V \to G \to G^{sa} \to 0,$$

with V a vector group and G^{sa} semi-abelian. We say that G is *completely non-trivial* (as a vector extension) if it does not have a direct factor $\mathbb{G}_a$. In other words, the classifying map

$$V^\vee \to \text{Ext}^1(G^{sa}, \mathbb{G}_a)$$

is injective; see Corollary 4.13.

Lemma 6.8. *Let G_1 be the Zariski closure of G_{tor} in G. Then G_1 is a completely non-trivial vector extension of G^{sa}. Moreover, there is a decomposition*

$$G \cong V_1 \times G_1$$

with a vector group V_1, i.e. G_1 is the maximal completely non-trivial subextension of G^{sa} contained in G.

Proof We have $G_{\text{tor}} \cong G_{\text{tor}}^{sa}$, hence the image of $G_1 \to G^{sa}$ contains all torsion points. They are dense in G^{sa}, hence $G_1 \to G^{sa}$ is surjective. This makes G_1 a vector extension of G^{sa}. By construction, $(G_1)_{\text{tor}}$ is dense in G_1. If it was not completely non-trivial, we would have a decomposition $G_1 = G_2 \times \mathbb{G}_a$ and the torsion points would not be dense.

Finally, let V be the vector part of G, $W = G_1 \cap V$ and choose a direct complement V_1 of W in V. The natural map

$$G_1 \times V_1 \to G$$

is an isomorphism. □

Proof of Theorem 6.3. By Lemma 6.8 we have

$$G \cong G_1 \times V_1$$

such that G_{tor} is dense in G_1 and V_1 is a vector group. By Lemma 6.5, the theorem holds for G_1.

By Lemma 6.6, there is a decomposition $V_1 \cong (V_1)_\Sigma \times V_{\text{tr}}$ such that ϕ is algebraic on $(V_1)_\Sigma$ and purely transcendental on V_{tr}. This completes the proof. □

The interplay between algebraic and transcendental morphisms is subtle. In the situation of Theorem 6.3 let

$$G \cong V_1 \times G_1, \quad G' \cong V_2 \times G_2$$

be decompositions of the algebraic groups G and G' into a vector group and a completely non-trivial vector extension of its semi-abelian part, as in Lemma 6.8. The analytic homomorphism ϕ decomposes as a (2×2)-matrix

$$\phi = \begin{pmatrix} \phi_{11} & \phi_{12} \\ \phi_{21} & \phi_{22} \end{pmatrix}$$

with $\phi_{11} \in \mathrm{Hom}(V_1^{\mathrm{an}}, V_2^{\mathrm{an}})$, $\phi_{12} \in \mathrm{Hom}(V_1^{\mathrm{an}}, G_2^{\mathrm{an}})$, $\phi_{21} \in \mathrm{Hom}(G_1^{\mathrm{an}}, V_2^{\mathrm{an}})$ and $\phi_{22} \in \mathrm{Hom}(G_1^{\mathrm{an}}, G_2^{\mathrm{an}})$.

Corollary 6.9. *In this situation, we have:*

1. *ϕ_{11} and ϕ_{22} are algebraic and defined over $\overline{\mathbb{Q}}$;*
2. *$\phi_{21} = 0$;*
3. *there is a decomposition $V_1 \cong V_{1,\mathrm{tr}} \times V_{1,\mathrm{alg}}$ such that the maps*

$$V_{1,\mathrm{tr}}^{\mathrm{an}} \to G_2^{sa,\mathrm{an}}, \quad V_{1,\mathrm{tr}}^{\mathrm{an}} \to G_2^{sa,\mathrm{an}}$$

induced by ϕ_{12} are purely transcendental and algebraic over $\overline{\mathbb{Q}}$, respectively.

Proof By the proof of Theorem 6.3, we have $G_1 \subset G_{\mathrm{alg}}$ and $V_{\mathrm{tr}} \subset V_1$. In particular, ϕ_{21} and ϕ_{22} are algebraic and defined over $\overline{\mathbb{Q}}$. If $\phi_{21} \colon G_1^{\mathrm{an}} \to V_2^{\mathrm{an}}$ was non-zero, we would be able to split off a factor $\mathbb{G}_a$ from G_1. This is impossible because G_1 is completely non-trivial.

Any analytic homomorphism $\phi_{11} \colon V_1^{\mathrm{an}} \to V_2^{\mathrm{an}}$ is algebraic over $\mathbb{C}$. It agrees with the analytification of the $\mathbb{C}$-linear map $\mathfrak{v}_{1,\mathbb{C}} \to \mathfrak{v}_{2,\mathbb{C}}$. By assumption it is induced by a $\overline{\mathbb{Q}}$-linear map $\mathfrak{v}_1 \to \mathfrak{v}_2$, hence it is even algebraic over $\overline{\mathbb{Q}}$.

We decompose V_1 as in Theorem 6.3 in this special case. Then ϕ_{12} is algebraic and defined over $\overline{\mathbb{Q}}$ on $V_{1,\mathrm{alg}}$, and purely transcendental on $V_{1,\mathrm{tr}}$. It remains to show that the composition $V_{1,\mathrm{tr}} \to G_2 \to G_2^{sa}$ is purely transcendental. In order to simplify notation, we write W for $V_{1,\mathrm{tr}}$ and G' for G_2. We apply Theorem 6.3 to $W \to G'^{sa}$. Accordingly there is a decomposition $W \cong W_{\mathrm{tr}} \times W_{\mathrm{alg}}$ such that the map is algebraic on W_{alg} and purely transcendental on W_{tr}. The algebraic map $W_{\mathrm{alg}} \to G'^{sa}$ vanishes because W_{alg} is a vector group and G'^{sa} is semi-abelian. This implies that we get an induced algebraic map $W_{\mathrm{alg}} \to V'$ where V' is the vector part of G'. This contradicts that $W_{\mathrm{alg}}^{\mathrm{an}} \to G'^{\mathrm{an}}$ is purely transcendental. We conclude that W_{alg} is in fact 0 and $W^{\mathrm{an}} \to G'^{sa,\mathrm{an}}$ is purely transcendental. $\square$

7

The Formalism of the Period Conjecture

The Period Conjecture predicts relations between the periods of algebraic varieties or, more generally, periods of motives. We explain the abstract set-up behind the explicit formulation. Our machinery will be applied mostly to periods of 1-motives, but also in a couple of other cases.

7.1 Periods

We first introduce periods and formal periods, following [HMS17, Definition 5.1.1] and [Hub20, Definition 3.6].

Throughout we fix subfields $K, L \subset \mathbb{C}$. Their compositum KL is the subfield generated by K and L. To simplify notation, we work under the hypothesis $K \cap L = \mathbb{Q}$.

Definitions 7.1. 1. Let (K, L)-Vect be the category of tuples $V = (V_K, V_L, \phi)$ where V_K and V_L are finite-dimensional vector spaces over K and L, respectively, and $\phi \colon V_K \otimes_K \mathbb{C} \to V_L \otimes_L \mathbb{C}$ is a $\mathbb{C}$-linear isomorphism. Morphisms are pairs of linear maps such that the diagram

$$
\begin{array}{ccc}
V_K \otimes_K \mathbb{C} & \xrightarrow{f_K \otimes_K \mathbb{C}} & W_K \otimes_K \mathbb{C} \\
\phi_V \downarrow & & \downarrow \phi_W \\
V_L \otimes_L \mathbb{C} & \xrightarrow{f_L \otimes_L \mathbb{C}} & W_L \otimes_L \mathbb{C}
\end{array}
$$

commutes.

2. Given $V \in (K, L)$-Vect, we define the *set of periods* of V as

$$
\mathcal{P}(V) = \mathrm{im}(V_K \times V_L^\vee \to \mathbb{C}), \quad (\sigma, \omega) \mapsto \omega_{\mathbb{C}}(\phi(\sigma_{\mathbb{C}}))
$$

and the *space of periods* $\mathcal{P}\langle V \rangle$ as the additive group generated by it. Here we write $\sigma_{\mathbb{C}}$ and $\omega_{\mathbb{C}}$ for the images of σ and ω in $V_K \otimes \mathbb{C}$ and $V_L^\vee \otimes \mathbb{C}$, respectively.

3. If $\mathcal{C}$ is a category, $V : \mathcal{C} \to (K, L)$-Vect a functor, we introduce

$$\mathcal{P}(\mathcal{C}) = \bigcup_{X \in \mathcal{C}} \mathcal{P}(V(X)).$$

Remarks 7.2. 1. The category (K, L)-Vect is $\mathbb{Q}$-linear and abelian. We could even turn it into a rigid tensor category and then $V \mapsto V_K$ becomes a so-called *fibre functor* if $K \subset L$. We are not going to use this fact.

2. The abelian group $\mathcal{P}\langle V \rangle$ is even a KL-vector space because of the bilinearity of the map $V_K \times V_L^\vee \to \mathbb{C}$. It has an alternative interpretation as the KL-vector space generated by the entries of the *period matrix*, the matrix of ϕ in a K-basis of V_K and an L-basis of V_L.

3. The set $\mathcal{P}(\mathcal{C})$ depends only on the objects in the image $V(\mathcal{C})$.

4. With $L = \mathbb{Q}$ this is the definition given in [HMS17] and [Hub20]. In the present monograph, the case $K = \mathbb{Q}$, $L = \overline{\mathbb{Q}}$ will be of most interest because 1-motives are a homological theory, whereas the other references take the cohomological point of view. In both cases we want to compare de Rham *cohomology* (the $\overline{\mathbb{Q}}$-component) with singular *homology* (the $\mathbb{Q}$-component).

5. We may replace the category $\mathcal{C}$ and the functor V by a diagram D (i.e. an oriented graph) and a representation V. Its periods are simply defined as the periods of the path category of D and the induced functor with values in (K, L)-Vect. This is the point of view taken originally by Nori and also in [HMS17]. It will play only a very minor role in our monograph, in the proof of Theorem 13.3 on the Period Conjecture for curves.

Example 7.3. The main case of interest for us is the category of iso-1-motives over $\overline{\mathbb{Q}}$; see Chapter 8 below. The functor V is given by the singular realisation, the de Rham realisation and by the period isomorphism.

Example 7.4. Given a short exact sequence

$$0 \to V_1 \to V \to V_2 \to 0$$

in (K, L)-Vect, the period matrix for V (in adapted bases) is upper block triangular, i.e. of the form

$$\begin{pmatrix} A & B \\ 0 & C \end{pmatrix},$$

such that A is the period matrix of V_1 and C the period matrix of V_2. In particular,

$$\mathcal{P}\langle V_1 \rangle + \mathcal{P}\langle V_2 \rangle \subset \mathcal{P}\langle V \rangle.$$

This is not an equality in general.

Lemma 7.5. *Let C be an additive category and $V \colon C \to (K, L)$-Vect an additive functor. Then $\mathcal{P}(C)$ is a KL-vector space. For $M \in (K, L)$-Vect we have*

$$\mathcal{P}\langle M \rangle = \mathcal{P}(\langle M \rangle),$$

where $\langle M \rangle \subset (K, L)$-Vect is the full abelian subcategory generated by M and closed under subquotients, i.e. the morphisms in $\langle M \rangle$ are the same as in (K, L)-Vect, and for $X \in \langle M \rangle$ and Y a subquotient of X in (K, L)-Vect, the object Y is also in $\langle M \rangle$.

Proof It suffices to show that $\mathcal{P}(C)$ is closed under addition. If α_1 is a period of X_1 and α_2 is a period of X_2, then $\alpha_1 + \alpha_2$ is a period of $X_1 \oplus X_2$.

As a consequence, the periods of the category $\langle V \rangle$ form an abelian group. They contain the periods of V, hence

$$\mathcal{P}\langle V \rangle \subset \mathcal{P}(\langle V \rangle).$$

For the converse inclusion, note that if V_1 is a subquotient of V_2, then $\mathcal{P}(V_1) \subset \mathcal{P}(V_2)$ by Example 7.4.

Moreover, $\mathcal{P}(M^n) \subset \mathcal{P}\langle M \rangle$. As all objects of $\langle V \rangle$ are subquotients of M^n for some n, this shows that $\mathcal{P}(W) \subset \mathcal{P}\langle V \rangle$ for all objects of $\langle V \rangle$. $\qquad\square$

There are obvious relations between the periods of a category C. They are encoded in a space of formal periods.

Definition 7.6. Let C be an additive category, $V \colon C \to (K, L)$-Vect be an additive functor. The *space of formal periods* $\widetilde{\mathcal{P}}(C)$ is the KL-vector space generated by symbols (σ, ω) for $\sigma \in V_K(X)$, $\omega \in V_L(X)^\vee$ for all objects X of C, with relations given by the following.

- (Bilinearity) For all objects X and $\sigma_1, \sigma_2 \in V_K(X)$, $\omega_1, \omega_2 \in V_L(X)^\vee$, $a_1, a_2 \in K$, $b_1, b_2 \in L$,

$$(a_1\sigma_1 + a_2\sigma_2, b_1\omega_1 + b_2\omega_2) = a_1 b_1 (\sigma_1, \omega_2) + \cdots + a_2 b_2 (\sigma_2, \omega_2).$$

- (Functoriality) For all morphisms $f \colon X \to Y$ and $\omega \in V_{\overline{\mathbb{Q}}}(Y)^\vee$, $\sigma \in V_{\mathbb{Q}}(X)$,

$$(f^*\omega, \sigma) = (\omega, f_*\sigma).$$

Equivalently, the vector space $\widetilde{\mathcal{P}}(\mathcal{C})$ can be characterised as the quotient space

$$\widetilde{\mathcal{P}}(\mathcal{C}) = \left(\bigoplus_{X \in \mathcal{C}} V_K(X) \otimes_{\mathbb{Q}} V_L(X)^{\vee}\right) \Big/ \text{ functoriality.}$$

The bilinearity relation is incorporated into the tensor product.

Remark 7.7. We could also apply the same definition to formal periods of a diagram D and a representation $V: D \to (K,L)$-Vect. This is the point of view taken in [HMS17]. The resulting space of formal periods agrees with the space of formal periods of the additive hull of the path category of D.

It is often useful to break $\mathcal{C}$ into smaller pieces.

Definition 7.8. Let $\mathcal{C}$ be an abelian category, X an object of $\mathcal{C}$. By $\langle X \rangle$ we denote the smallest full subcategory of $\mathcal{C}$ that contains X and is closed under subquotients.

We have $\mathcal{P}\langle X \rangle = \mathcal{P}(\langle X \rangle)$ and this shows that this is the right category if we are trying to understand linear relations between periods of X.

Lemma 7.9. *For an additive functor* $\mathcal{C} = \langle X \rangle \to (K,L)$-*Vect, the elements of* $V_K(X) \otimes_{\mathbb{Q}} V_L(X)^{\vee}$ *generate* $\widetilde{\mathcal{P}}(\mathcal{C})$ *as a KL-vector space. In particular,*

$$\dim_{KL} \widetilde{\mathcal{P}}(\mathcal{C}) \leq (\dim_K V_K(X))^2.$$

Proof We need to verify that all generators of $\widetilde{\mathcal{P}}(\mathcal{C})$ can be expressed in terms of elements of $V_K(X) \otimes_{\mathbb{Q}} V_L(X)^{\vee}$.

If $f: Y \to Y'$ is a surjective morphism in $\mathcal{C}$, then all elementary tensors in $V_K(Y') \otimes_{\mathbb{Q}} V_L(Y')^{\vee}$ can be identified with elementary tensors in the tensor product $V_K(Y) \otimes_{\mathbb{Q}} V_L(Y)^{\vee}$ because $V_K(Y) \to V_K(Y')$ is surjective, i.e. every element in $V_K(Y')$ has the form $f_* \sigma$ for some $\sigma \in V_K(Y)$, and by consequence,

$$f_* \sigma \otimes \omega = \sigma \otimes f^* \omega \in \widetilde{\mathcal{P}}(\mathcal{C}).$$

In the same way, if $f: Y \to Y'$ is injective, then the elementary tensors on Y can be identified with some elementary tensors on Y' because $V_L(Y') \to V_L(Y)$ is surjective.

By assumption, every object of $\langle X \rangle$ is a subquotient of some X^n for $n \geq 1$. Hence it suffices to consider elementary tensors on X^n. Note that $V_K(X^n) \cong V_K(X)^n$ and that we can write an elementary tensor $\sigma \otimes \omega \in V_K(X^n) \otimes_{\mathbb{Q}} V_L(X^n)$ as

$$\sigma \otimes \omega = \sum_{k=1}^{n} (i_k)_* \sigma_k \otimes \omega,$$

where $\sigma_1, \ldots, \sigma_n$ are the components of σ. By the functoriality relation this yields the identification of

$$\sigma \otimes \omega = \sum_{k=1}^{n} \sigma_k \otimes i_k^* \omega$$

with an element of $V_K(X) \otimes_{\mathbb{Q}} V_L(X)^{\vee}$. □

Following Hörmann in [Hör21], there is an interesting alternative description of the space of relations in the abelian case. It is closer to the shape in which they will appear in the context of 1-motives and was motivated by it.

Given a short exact sequence

$$0 \to X_1 \overset{i}{\to} X^n \overset{p}{\to} X_2 \to 0$$

in a $\mathbb{Q}$-linear abelian category $\mathcal{C}$ and elements $(\sigma_1, \ldots, \sigma_n) \in i_*(V_K(X_1))$, $(\omega_1, \ldots, \omega_n) \in p^*(V_L(X_2)^{\vee})$, the functoriality relation implies that $\sum_{i=1}^{n} \sigma_i \otimes \omega_i$ vanishes in $\widetilde{\mathcal{P}}\langle X \rangle$. Actually, even the converse is true.

Proposition 7.10 (Hörmann [Hör21]). *For an additive functor* $\mathcal{C} = \langle X \rangle \to (K, L)$-*Vect, an element of the form* $\sum_{i=1}^{n} \sigma_i \otimes \omega_i$ *is in the kernel of the map* $V_K(X) \otimes_{\mathbb{Q}} V_L(X)^{\vee} \to \widetilde{\mathcal{P}}\langle X \rangle$ *if and only if there is a short exact sequence*

$$0 \to X_1 \overset{i}{\to} X^n \overset{p}{\to} X_2 \to 0$$

with $(\sigma_1, \ldots, \sigma_n) \in i_*(V_K(X_1))$, $(\omega_1, \ldots, \omega_n) \in p^*(V_L(X_2)^{\vee})$.

We omit the proof as we are not going to need this fact.

By construction, the space of formal periods comes with an evaluation map to $\mathbb{C}$.

Definition 7.11. Let $\mathcal{C}$ be an additive category and $V \colon \mathcal{C} \to (K, L)$-Vect an additive functor. We define the *evaluation map*

$$\mathrm{ev} \colon \widetilde{\mathcal{P}}(\mathcal{C}) \to \mathbb{C}$$

on (σ, ω) for σ in $V_K(X)$ and ω in $V_L(X)^{\vee}$ by

$$(\sigma, \omega) \mapsto \omega_{\mathbb{C}}(\phi(\sigma_{\mathbb{C}})).$$

The map is obviously well defined, KL-linear and has image $\mathcal{P}(\mathcal{C})$.

Definition 7.12. We define the *external duality functor*

$$\cdot^{\vee} \colon (K, L)\text{-Vect} \to (L, K)\text{-Vect}$$

by assigning the triple (V_K, V_L, ϕ) to $(V_L^{\vee}, V_K^{\vee}, \phi^{\vee})$.

This functor should not be confused with the internal duality functor on (K, L)-Vect, which maps (V_K, V_L, ϕ) to $(V_K^\vee, V_L^\vee, (\phi^\vee)^{-1})$.

Lemma 7.13. *Let $V: C \to (K, L)$-Vect be an additive functor. Then the period spaces $\mathcal{P}(C)$ and $\widetilde{\mathcal{P}}(C)$ do not change when we apply the external duality functor.*

Proof Let $X \in C$. The definition of $\widetilde{\mathcal{P}}(C)$ via V uses $V_K(X) \otimes_{\mathbb{Q}} V_L(X)^\vee$, whereas the definition via $\cdot^\vee \circ V$ uses $V_L^\vee \otimes_{\mathbb{Q}} (V_K^\vee)^\vee$. These spaces are identified by exchanging the factors. The compatibility with the evaluation map is the very definition of the dual $\phi^\vee$ of ϕ. $\qquad\square$

Remark 7.14. In the case of internal duality, we get the same statement for $\widetilde{\mathcal{P}}(C)$, but no longer for actual periods. External duality maps a period matrix to its transpose, so the period space remains the same. In contrast, internal duality maps the period matrix to the inverse of the transpose, hence the period space is divided by the determinant, and its periods are divided by the determinant.

7.2 The Period Conjecture

The Period Conjecture asserts that in certain cases the obvious relations are the only ones. We follow [Hub20]. As in the previous section, we fix subfields $K, L, K \cap L = \mathbb{Q}$. The cases of interest for the Period Conjecture are $K = \mathbb{Q}$, $L \subset \overline{\mathbb{Q}}$, or conversely.

Definition 7.15 (Huber [Hub20, Definition 3.7])**.** Let C be an additive category and $V: C \to (K, L)$-Vect an additive functor. We say that *the Period Conjecture holds for C* if the evaluation map $\widetilde{\mathcal{P}}(C) \to \mathcal{P}(C)$ is injective.

Remark 7.16. If C is the category of all Nori motives over $\overline{\mathbb{Q}}$ (see Appendix A), then this is the Period Conjecture as formulated by Kontsevich in [Kon99]. We refer the reader to [HMS17, Part III] for a detailed discussion. The conjecture is proved for the category of iso-1-motives in Theorem 9.10. Note that the above statement does not mention the tensor structure, which exists on the category of all motives (but not on 1-Mot$_{\overline{\mathbb{Q}}}$). We refer the reader to [Hub20] for the discussion of tensor products and the comparison of the above conjecture with Grothendieck's version predicting the transcendence degree of the algebra generated by the periods of a motive. The latter does not a play a role in our monograph.

Note that the space of formal periods $\widetilde{\mathcal{P}}(C)$ and hence the Period Conjecture depend only on the image of C under V. Hence we may assume without loss of generality that V is faithful.

Proposition 7.17 (Huber [Hub20, Proposition 5.2]). *Let* $F: \mathcal{C}' \to \mathcal{C}$ *and* $V: \mathcal{C} \to (K, L)$-Vect *be faithful exact functors between* $\mathbb{Q}$-*linear abelian categories. Then*

$$\widetilde{\mathcal{P}}(\mathcal{C}') \to \widetilde{\mathcal{P}}(\mathcal{C})$$

is injective if and only if F is full with image closed under taking subquotients.

Remark 7.18. The proof of this general fact relies on Nori's description of such categories as categories of comodules for an explicit coalgebra. In the cases of interest for us, we will give a direct proof in later chapters.

Corollary 7.19 (Fullness: Huber [Hub20, Corollary 5.3]). *Let $\mathcal{C}$ be a* $\mathbb{Q}$-*linear abelian category and* $V: \mathcal{C} \to (K, L)$-Vect *a faithful exact functor. If the Period Conjecture holds for $\mathcal{C}$, then V is full with image closed under taking subquotients.*

Proof Let $\bar{\mathcal{C}}$ be the full subcategory of (K, L)-Vect closed under taking subquotients generated by $\mathcal{C}$. Then the evaluation map factors as

$$\widetilde{\mathcal{P}}(\mathcal{C}) \to \widetilde{\mathcal{P}}(\bar{\mathcal{C}}) \to \mathbb{C}.$$

If the composition map is injective, so is the first map. By applying Proposition 7.17 to $F = V$, we deduce that V is full, and as a consequence $\mathcal{C}$ is equivalent to $\bar{\mathcal{C}}$. □

The Period Conjecture for an abelian category $\mathcal{C}$ can be broken down into parts. Recall from Definition 7.8 the subcategory $\langle X \rangle$ generated by a single object. We have $\mathcal{P}\langle X \rangle = \mathcal{P}(\langle X \rangle)$, so this is the right category if we want to understand linear relations between periods of X.

Lemma 7.20 (Huber [Hub20, Proposition 5.6]). *Let $\mathcal{C}$ be an abelian category and $V: \mathcal{C} \to (K, L)$-Vect a faithful exact functor. Then the following statements are equivalent:*

1. *The Period Conjecture holds for $\mathcal{C}$.*
2. *The Period Conjecture holds for $\langle X \rangle$ for all objects X of $\mathcal{C}$.*

Proof By Proposition 7.17 applied to $\langle X \rangle \to \mathcal{C}$, the natural map

$$\widetilde{\mathcal{P}}(\langle X \rangle) \to \widetilde{\mathcal{P}}(\mathcal{C})$$

is injective. If $\widetilde{\mathcal{P}}(\mathcal{C}) \to \mathbb{C}$ is injective, so is the composition

$$\widetilde{\mathcal{P}}(\langle X \rangle) \to, \widetilde{\mathcal{P}}(\mathcal{C}) \to \mathbb{C}$$

for every object X. This shows that statement (1) implies statement (2). Conversely, we have

$$\mathcal{C} = \bigcup_{X \in \mathcal{C}} \langle X \rangle$$

because a morphism $f\colon X \to Y$ in our abelian category $\mathcal{C}$ is already a morphism in the subcategory $\langle X \oplus Y \rangle$. As a consequence we have

$$\widetilde{\mathcal{P}}(\mathcal{C}) = \varinjlim_{X \in \mathcal{C}} \widetilde{\mathcal{P}}(\langle X \rangle).$$

If the evaluation map is injective for every X, it is injective on the inductive limit. $\qquad\square$

Using Hörmann's alternative description of the space of formal periods, we can reformulate the conjecture.

Corollary 7.21. *The Period Conjecture holds for the abelian category $\mathcal{C} = \langle X \rangle$ if and only if for an element $\sum_{i=1}^{n} \sigma_i \otimes \omega_i \in V_K(X) \otimes_{\mathbb{Q}} V_L(X)^{\vee}$ in the kernel of the evaluation map there is a short exact sequence*

$$0 \to X_1 \xrightarrow{i} X^n \xrightarrow{p} X_2 \to 0$$

with $(\sigma_1, \ldots, \sigma_n) \in i_*(V_K(X_1))$, $(\omega_1, \ldots, \omega_n) \in p^*(V_L(X_2)^{\vee})$.

Proof Apply Proposition 7.10. $\qquad\square$

The advantage of taking the subcategory $\langle X \rangle$ for an individual $X \in \mathcal{C}$ is that its period space is finite dimensional over KL; indeed its dimension is bounded by $\dim_K V_K(X)^2$. Hence it makes sense to ask what the dimension actually is.

Corollary 7.22. *Let $\mathcal{C}$ be a $\mathbb{Q}$-linear additive category and X an object of $\mathcal{C}$. Then the Period Conjecture holds for $\langle X \rangle$ if and only if*

$$\dim_{KL} \widetilde{\mathcal{P}}(\langle X \rangle) = \dim_{KL} \mathcal{P}\langle X \rangle.$$

Proof The evaluation map $\widetilde{\mathcal{P}}(\langle X \rangle) \to \mathcal{P}\langle X \rangle$ for $\langle X \rangle$ is surjective. Hence it is injective if and only if it is an isomorphism and if and only the dimensions of the two finite-dimensional vector spaces agree. $\qquad\square$

It remains for us to understand the dimension of $\widetilde{\mathcal{P}}(\langle X \rangle)$. This question is answered via Nori's version of Tannaka theory without a tensor product. We recall the main player.

We keep concentrating on the case of an abelian category generated by a single object X, in the sense of $\mathcal{C} = \langle X \rangle$.

Definition 7.23. Let $T\colon \mathcal{C} = \langle X \rangle \to \mathbb{Q}\text{-Vect}$ be a faithful exact functor. We introduce the spaces

$$\mathrm{End}(T) = \left\{ (f_Y) \in \prod_{Y \in \mathcal{C}} \mathrm{End}_{\mathbb{Q}}(T(Y)) \mid \forall g\colon Y \to Y'\colon f_{Y'} \circ T(g) = T(g) \circ f_Y \right\}$$

and

$$\mathcal{A}(\mathcal{C}, T) = \mathrm{End}(T)^{\vee}.$$

Lemma 7.24. *If* $C = \langle X \rangle$, *then elements* (f_Y) *of* $\mathrm{End}(T)$ *are uniquely determined by* f_X. *In other words,* $\mathrm{End}(T) \subset \mathrm{End}_{\mathbb{Q}}(T(X))$ *and in consequence*

$$\dim_{\mathbb{Q}} \mathrm{End}(T) \leq (\dim_{\mathbb{Q}} T(X))^2.$$

In particular, the dimension is finite.

Proof   The argument is the same as in the proof of Lemma 7.9, phrased in a slightly different language.

Let $f = (f_Y)$ be in $\mathrm{End}(T)$. Assume that $g \colon Y \to Y'$ is surjective in $\langle X \rangle$, then so is $T(g) \colon T(Y) \to T(Y')$. From the commutative diagram

$$
\begin{array}{ccc}
T(Y) & \xrightarrow{\;T(g)\;} & T(Y') \\
{\scriptstyle f_Y}\downarrow & & \downarrow{\scriptstyle f_{Y'}} \\
T(Y) & \xrightarrow{\;T(g)\;} & T(Y'),
\end{array}
$$

we deduce that $f_{Y'}$ is uniquely determined by f_Y. In the same way, if $g \colon Y \to Y'$ is injective, then f_Y is uniquely determined by $f_{Y'}$. By assumption, every object of $\langle X \rangle$ is a subquotient of M^n for some $n \geq 1$. Hence f is determined by the components f_{X^n}. We write $f_{X^n} \in \mathrm{End}(T(X)^n)$ as a matrix ϕ_{ij} with entries in $\mathrm{End}(T(X))$. The entry ϕ_{ij} is the composition

$$T(X) \xrightarrow{T(\iota_i)} T(X)^n \xrightarrow{f_{X^n}} T(X)^n \xrightarrow{T(p_j)} T(X)$$

with the injection $\iota_i \colon X \to X^n$ and the projections $p_j \colon X^n \to X$. We have commutative diagrams

$$
\begin{array}{ccccc}
T(X) & \xrightarrow{\;T(\iota_i)\;} & T(X)^n & \xrightarrow{\;T(p_j)\;} & T(X) \\
{\scriptstyle f_X}\downarrow & & {\scriptstyle f_{X^n}}\downarrow & & \downarrow{\scriptstyle f_X} \\
T(X) & \xrightarrow{\;T(\iota_i)\;} & T(X)^n & \xrightarrow{\;T(p_j)\;} & T(X).
\end{array}
$$

The map from the top left to the bottom right is ϕ_{ij}. For $i \neq j$, the composition $p_j \circ \iota_i$ vanishes, hence so does ϕ_{ij}. For $i = j$ the composition is the identity, hence $\phi_{ii} = f_X$. In particular, the map f_{X^n} is uniquely determined by f_X. This finishes the proof of the first claim. The others follows directly from this fact. □

The set $\mathrm{End}(T)$ is stable under composition, making it into a unital $\mathbb{Q}$-algebra. Dually, $\mathcal{A}(\mathcal{C}, T)$ is a *counital coalgebra*: it is equipped with a *comultiplication*, i.e. a $\mathbb{Q}$-linear map

$$\mathcal{A}(\mathcal{C}, T) \to \mathcal{A}(\mathcal{C}, T) \otimes_{\mathbb{Q}} \mathcal{A}(\mathcal{C}, T)$$

satisfying axioms dual to the axioms of a unital algebra.

For every $Y \in \langle X \rangle$, the vector space $T(Y)$ has a natural action of $\mathrm{End}(T)$ where an element f operates via its f_Y-component. This defines a functor

$$\widetilde{T} \colon \langle X \rangle \to \mathrm{End}(T)\text{-Mod},$$

where $\mathrm{End}(T)$-Mod denotes the category of finitely generated $\mathrm{End}(T)$-modules, or equivalently $\mathrm{End}(T)$-modules whose underlying $\mathbb{Q}$-vector space is finite dimensional. By adjunction, the structure map

$$\mathrm{End}(T) \otimes V \to V$$

of an $\mathrm{End}(T)$-module V induces a $\mathbb{Q}$-linear map

$$V \to \mathcal{A}(\langle X \rangle, T) \otimes_{\mathbb{Q}} V,$$

turning V into a *comodule* whose underlying $\mathbb{Q}$-vector space is finite dimensional. For the axioms of a comodule, see for example [HMS17, Section 7.5.2]. They are dual to the axioms of a module under an algebra.

The significance of $\mathrm{End}(T)$ and $\mathcal{A}(\mathcal{C}, T)$ is due to the following strong property.

Proposition 7.25. *Let* $T \colon \mathcal{C} = \langle X \rangle \to \mathbb{Q}\text{-Vect}$ *be a faithful and exact functor. Then $\mathcal{C}$ is equivalent to the category of finite-dimensional $\mathrm{End}(T)$-modules, or, equivalently to the category of finite-dimensional $\mathbb{Q}$-vector spaces equipped with the structure of an $\mathcal{A}(\mathcal{C}, T)$-comodule.*

Remark 7.26. We are not going to use this structural result in our applications to the Period Conjecture. We defer the deduction from the existing literature to the end of the chapter.

For later use, we give a more explicit description of $\mathcal{A}(\mathcal{C}, T)$.

Lemma 7.27. *In the situation of Definition 7.23, we have*

$$\mathcal{A}(\mathcal{C}, T) = \left(\bigoplus_{Y \in \mathcal{C}} \mathrm{End}_{\mathbb{Q}}(T(Y))^{\vee} \right) \Big/ \text{functoriality}.$$

The functoriality relations are generated by elements of the form $\sigma \otimes f^*(\omega) - f_*(\sigma) \otimes \omega$ for all $f \colon Y \to Y'$ in $\mathcal{C}$ and $\sigma \in T(Y)$, $\omega \in T(Y')^{\vee}$ under the identification $\mathrm{End}(T(Y))^{\vee} \cong T(Y) \otimes T(Y)^{\vee}$.

Proof Let $\mathcal{A}'(\mathcal{C}, T)$ be the object on the right-hand side. By definition, $\mathcal{A}'(\mathcal{C}, T)^{\vee} \cong \mathrm{End}(T)$ (the direct sum turns into a product and the quotient into a subobject). Taking duals again, we get

$$(\mathcal{A}'(\mathcal{C}, T)^{\vee})^{\vee} \cong \mathrm{End}(T)^{\vee} = \mathcal{A}(\mathcal{C}, T).$$

As the vector spaces are finite dimensional, they are isomorphic to their double duals. Hence we have shown that

$$\mathcal{A}'(\mathcal{C}, T) \cong \mathcal{A}(\mathcal{C}, T).$$ □

The coalgebra point of view has the advantage of generalising to all $\mathcal{C}$.

Definition 7.28. Let $\mathcal{C}$ be a $\mathbb{Q}$-linear abelian category and $T \colon \mathcal{C} \to \mathbb{Q}$-Vect a faithful exact functor into the category of finite-dimensional $\mathbb{Q}$-vector spaces. We put

$$\mathcal{A}(\mathcal{C}, T) = \left(\bigoplus_{Y \in \mathcal{C}} \mathrm{End}_{\mathbb{Q}}(T(Y))^{\vee} \right) \Big/ \text{ functoriality},$$

with functoriality interpreted as in Lemma 7.27.

As in the case where $\mathcal{C}$ is generated by a single object, this vector space has a natural $\mathbb{Q}$-coalgebra structure and every $T(Y)$ inherits the structure of a comodule.

Theorem 7.29 (Nori: see [HMS17, Chapter 7]). *Let $\mathcal{C}$ be a $\mathbb{Q}$-linear abelian category and $T \colon \mathcal{C} \to \mathbb{Q}$-Vect a faithful exact functor into the category of finite-dimensional $\mathbb{Q}$-vector spaces. Then $\mathcal{C}$ is equivalent to the category of $\mathcal{A}(\mathcal{C}, T)$-comodules whose underlying $\mathbb{Q}$-vector spaces are finite dimensional.*

The proof of this theorem is non-trivial. Proposition 7.25 is a key step in its proof. In our monograph we use the opposite approach and deduce Proposition 7.25 from Theorem 7.29 instead.

Proof of Proposition 7.25. We apply the theorem to $\mathcal{C} = \langle X \rangle$. By Lemma 7.27, the coalgebra $\mathcal{A}(\mathcal{C}, T)$ is dual to the algebra $\mathrm{End}(T)$. It is finite dimensional by Lemma 7.24, hence the category of finitely generated $\mathrm{End}(T)$-modules is equivalent to the category of $\mathcal{A}(\mathcal{C}, T)$-comodules that are finite dimensional over $\mathbb{Q}$.   □

Remark 7.30. We have decided to deduce Proposition 7.25 from Theorem 7.29, but actually the converse is also true and implicitly this is the way Theorem 7.29 is shown in [HMS17]. The argument in [HMS17] is made more complicated by allowing other base rings than $\mathbb{Q}$. At heart, the result is even older. It appears as a step in the proof of Tannaka duality in [DM82]. Let us explain how Theorem 7.29 relates to Tannaka duality, even if these issues are not relevant for the rest of the monograph.

If $\mathcal{C}$ is not only abelian, but a so-called Tannaka category with fibre functor $T \colon \mathcal{C} \to \mathbb{Q}$-Vect (i.e. equipped with with a unitary commutative associative tensor product and T being a faithful tensor functor), then the coalgebra

$\mathcal{A}(\mathcal{C}, T)$ is endowed with a commutative multiplication, making it into a Hopf algebra. It follows that $G = \mathrm{Spec}(\mathcal{A}(\mathcal{C}, T))$ is a group scheme. There is an equivalence of categories between $\mathcal{A}(\mathcal{C}, T)$-comodules and representations of G, identifying $\mathcal{C}$ (as a Tannakian category) with finite-dimensional representations of G.

Having introduced $\mathrm{End}(T)$ and $\mathcal{A}(\mathcal{C}, T)$ and explained their significance, we can now establish the dimension formula for the space of abstract periods that we are after. As before, let L be a subfield of $\mathbb{C}$.

Proposition 7.31. *Let $\mathcal{C} = \langle X \rangle$ be a $\mathbb{Q}$-linear additive category and $V \colon \mathcal{C} \to (\mathbb{Q}, L)$-Vect a faithful and exact additive functor. Then*

$$\dim_L \widetilde{\mathcal{P}}(\mathcal{C}) = \dim_{\mathbb{Q}} \mathcal{A}(\mathcal{C}, V_{\mathbb{Q}}) = \dim_{\mathbb{Q}} \mathrm{End}(V_{\mathbb{Q}}).$$

Proof The second equality holds because the vector spaces are $\mathbb{Q}$-dual to each other.

The definitions of $\widetilde{\mathcal{P}}(\mathcal{C})$ and $\mathcal{A}(\mathcal{C}, V_{\mathbb{Q}})$ are very similar. They become isomorphic after base change to $\mathbb{C}$; in particular, they have the same dimensions. In detail:

$$\mathcal{A}(\mathcal{C}, V_{\mathbb{Q}}) \otimes_{\mathbb{Q}} \mathbb{C} = \left(\bigoplus_{Y \in \mathcal{C}} \mathrm{End}_{\mathbb{C}}(V_{\mathbb{C}}(Y))^{\vee} \right) \Big/ \text{functoriality}$$

$$= \left(\bigoplus_{Y \in \mathcal{C}} V_{\mathbb{C}}(Y) \otimes_{\mathbb{C}} V_{\mathbb{C}}(Y)^{\vee} \right) \Big/ \text{functoriality},$$

where we write $V_{\mathbb{C}} := V_{\mathbb{Q}} \otimes_{\mathbb{Q}} \mathbb{C}$. On the other hand,

$$\widetilde{\mathcal{P}}(\mathcal{C}) \otimes_L \mathbb{C} = \left(\bigoplus_{Y \in \mathcal{C}} V_{\mathbb{Q}}(Y) \otimes_{\mathbb{Q}} V_L(Y)^{\vee} \otimes_{\overline{\mathbb{Q}}} \mathbb{C} \right) \Big/ \text{functoriality}$$

$$= \left(\bigoplus_{Y \in \mathcal{C}} V_{\mathbb{C}}(Y) \otimes_{\mathbb{C}} (V_L(Y)^{\vee} \otimes_L \mathbb{C}) \right) \Big/ \text{functoriality}$$

$$= \left(\bigoplus_{Y \in \mathcal{C}} V_{\mathbb{C}}(X) \otimes_{\mathbb{C}} (V_{\mathbb{C}}(Y)^{\vee} \mathbb{C}) \right) \Big/ \text{functoriality}$$

because $V_L \otimes_L \mathbb{C} \cong V_{\mathbb{C}}$. $\qquad\qquad\qquad\qquad\qquad\qquad\qquad\qquad\square$

In the language of [Hub20], $\widetilde{\mathcal{P}}(\mathcal{C})$ is a semi-torsor under $\mathcal{A}(\mathcal{C}, V_{\mathbb{Q}})$.

Corollary 7.32. *Let $\mathcal{C}$ be a $\mathbb{Q}$-linear abelian category, X an object of $\mathcal{C}$ and $V \colon \mathcal{C} \to (\mathbb{Q}, L)$-Vect a faithful and exact additive functor. Then the Period Conjecture holds for $\langle X \rangle$ if and only if*

$$\dim_L \mathcal{P}\langle X \rangle = \dim_{\mathbb{Q}} \mathrm{End}(V_{\mathbb{Q}}|_{\langle X \rangle}).$$

PART TWO

PERIODS OF DELIGNE 1-MOTIVES

8

Deligne's 1-Motives

In this chapter we review Deligne's category of 1-motives over an algebraically closed field k embedded into $\mathbb{C}$. The cases of interest to us are $k = \mathbb{C}$ and $k = \overline{\mathbb{Q}}$. In the latter case, Section 8.2 and to a greater extent Section 8.3 contain new results.

8.1 The Category and the Realisation Functors

Definition 8.1 (Deligne [Del74, Chapter 10]). A 1-*motive* $M = [L \to G]$ over k is the datum given by a semi-abelian group G over k, a free abelian group L of finite rank and a group homomorphism $L \to G(k)$. Morphisms of 1-motives are morphisms of complexes $L \to G$. The category 1-Mot$_k$ of *iso-1-motives* has the same objects, but morphism tensored by $\mathbb{Q}$.

Remark 8.2. The category of iso-1-motives is abelian. In this monograph, we are working in the category of iso-1-motives. The arguments often involve replacing a 1-motive $[L \to G]$ by an isogenous 1-motive $[L' \to G']$. This will sometimes happen tacitly.

We need to spell out the singular and de Rham realisation defined in [Del74, Chapter 10] in detail.

8.1.1 The Singular Realisation

Let $M = [L \xrightarrow{u} G]$ be a 1-motive. We write G^{an} for the commutative Lie group over $\mathbb{C}$ attached to G. The associated exponential sequence is

$$0 \to H_1^{\mathrm{sing}}(G^{\mathrm{an}}, \mathbb{Z}) \to \mathrm{Lie}(G^{\mathrm{an}}) \xrightarrow{\exp} G^{\mathrm{an}} \to 0,$$

where we have used the identification made in Section 5.3, Equation (5.4).

Definition 8.3. Let $T_{\text{sing}}(M)$ be the fibre product of L and $\text{Lie}(G^{\text{an}})$ over G^{an} under the structure map $u \colon L \to G^{\text{an}}$ and the exponential map $\exp$

$$
\begin{CD}
T_{\text{sing}}(M) @>>> \text{Lie}(G^{\text{an}}) \\
@VVV @VV{\exp}V \\
L @>{u}>> G.
\end{CD}
$$

The vector space $V_{\text{sing}}(M) = T_{\text{sing}}(M) \otimes \mathbb{Q}$ is called the *singular realisation* of M.

By construction, there is a short exact sequence

$$
0 \to H_1(G^{\text{an}}, \mathbb{Q}) \to V_{\text{sing}}(M) \to L \otimes \mathbb{Q} \to 0.
$$

In particular, this gives $V_{\text{sing}}(M) \cong H_1(G^{\text{an}}, \mathbb{Q})$ if $L = 0$, and $V_{\text{sing}}(M) = L_{\mathbb{Q}}$ if G is trivial. The vector space $V_{\text{sing}}(M)$ carries a weight filtration with

$$
W_n V_{\text{sing}}(M) = \begin{cases} 0 & n \leq -3, \\ H_1(T, \mathbb{Q}) & n = -2, \\ H_1(G, \mathbb{Q}) & n = -1, \\ V_{\text{sing}}(M) & n \geq 0. \end{cases}
$$

Here $0 \to T \to G \to A \to 0$ is the decomposition of G into a torus and an abelian variety.

Lemma 8.4. *The functor* $V_{\text{sing}} \colon 1\text{-Mot}_k \to \mathbb{Q}\text{-Vect}$ *is faithful and exact.*

Proof Exactness holds because taking a Lie algebra and taking a pullback are exact functors. In order to verify faithfulness it suffices to show that $V_{\text{sing}}(M) = 0$ implies $M = 0$ in 1-Mot_k. Hence we consider some M with $V_{\text{sing}}(M) = 0$. This implies $H_1(G, \mathbb{Q}) = 0$, and in consequence G is trivial and $V_{\text{sing}}(M) = L_{\mathbb{Q}}$. In this case L also has to be trivial.   □

8.1.2 The Universal Vector Extension

As in Chapter 4, let $\mathcal{G}$ be the category of connected commutative algebraic groups. We denote by $\mathcal{G}_{\mathbb{Q}}$ its isogeny category. We enlarge the category 1-Mot_k to a bigger abelian category, which we call 1-MOT_k. Its objects are of the form $[L \to G]$ with G in the category $\mathcal{G}$, and its morphisms are given by morphisms of complexes tensored by $\mathbb{Q}$. The category $\mathcal{G}_{\mathbb{Q}}$ can be identified with a full subcategory of 1-MOT_k by $G \mapsto [0 \to G]$. It is closed under extensions. This means that the bifunctor Ext^1 in $\mathcal{G}_{\mathbb{Q}}$ is the same as the bifunctor Ext^1 in 1-MOT_k restricted to $\mathcal{G}_{\mathbb{Q}}$.

We briefly recall the universal vector extension used in Construction 10.1.7 of [Del74]. In Section 4.4 we introduced the universal vector extension

$$0 \to \mathrm{Ext}^1(G, \mathbb{G}_a)^\vee \to G^\natural \to G \to 0.$$

It depended on the computation of $\mathrm{Ext}^1_G(G, \mathbb{G}_a) = \mathrm{Ext}^1_{\mathcal{G}_\mathbb{Q}}(G, \mathbb{G}_a)$ for the additive group $\mathbb{G}_a$.

Lemma 8.5. *For* $M = [L \to G]$ *in* 1-Mot_k *there is a natural short exact sequence of k-vector spaces*

$$0 \to \mathrm{Hom}_{\mathrm{ab}}(L, \mathbb{G}_a) \to \mathrm{Ext}^1_{1-\mathrm{MOT}}(M, \mathbb{G}_a) \to \mathrm{Ext}^1_G(G, \mathbb{G}_a) \to 0.$$

In particular, they are finite-dimensional vector spaces.

Proof An application of the functor $\mathrm{Hom}(-, [0 \to \mathbb{G}_a])$ to the short exact sequence

$$0 \to [0 \to G] \to M \to [L \to 0] \to 0$$

in 1-MOT_k and identifying the 1-motives $[0 \to G]$ and $[0 \to \mathbb{G}_a]$ with G and $\mathbb{G}_a$, respectively, by our convention above, induces a long exact sequence

$$\mathrm{Hom}(G, \mathbb{G}_a) \to \mathrm{Ext}^1([L \to 0], \mathbb{G}_a) \to \mathrm{Ext}^1(M, \mathbb{G}_a) \to \mathrm{Ext}^1(G, \mathbb{G}_a). \quad (8.1)$$

The first term vanishes because G is semi-abelian. Given an extension

$$0 \to \mathbb{G}_a \to E \to G \to 0$$

in $\mathrm{Ext}^1(G, \mathbb{G}_a)$, we can lift the structure map $L \to G$ to $L \to E$ because L is free. We obtain a short exact sequence

$$0 \to [0 \to \mathbb{G}_a] \to [L \to E] \to [L \to G] \to 0,$$

which is in $\mathrm{Ext}^1_{1-\mathrm{MOT}_k}(M, \mathbb{G}_a)$ because $[L \to G] = M$. This makes the last map of (8.1) surjective.

Elements of $\mathrm{Ext}^1([L \to 0], \mathbb{G}_a)$ are of the form

$$0 \to [0 \to \mathbb{G}_a] \to [L \to \mathbb{G}_a] \to [L \to 0] \to 0,$$

hence they can be identified with homomorphisms $L \to \mathbb{G}_a$.

The group $\mathrm{Ext}^1(G, \mathbb{G}_a)$ is finite dimensional by Corollary 4.10. On the other hand, $\dim \mathrm{Hom}(L, \mathbb{G}_a) = \mathrm{rk}(L)$ and the last statement follows. $\square$

Our construction can be easily extended to the case when the group $\mathbb{G}_a$ is replaced by a finite-dimensional vector space V.

Definition 8.6. Let $M = [L \to G]$ be in 1-Mot$_k$. A *vector extension* of M is an extension of the form

$$0 \to [0 \to V] \to [L \to G'] \to [L \to G] \to 0$$

for a vector group V.

By the same arguments as in the case of semi-abelian varieties in Section 4.4, the datum of a vector extension of M is equivalent to the datum of a classifying map

$$\mathrm{Ext}^1(M, \mathbb{G}_a)^\vee \to V.$$

The identity map is a distinguished choice of classifying map for $V = \mathrm{Ext}^1(M, \mathbb{G}_a)^\vee$.

Definition 8.7. For a 1-motive $M = [L \to G]$ in 1-Mot$_k$, we call the extension

$$0 \to \mathrm{Ext}^1(M, \mathbb{G}_a) \to [L \to M^\natural] \to [L \to G] \to 0$$

in $1-\mathrm{MOT}_k$ corresponding to the classifying map

$$\mathrm{id} \colon \mathrm{Ext}^1(M, \mathbb{G}_a)^\vee \to \mathrm{Ext}^1(M, \mathbb{G}_a)^\vee$$

the *universal vector extension* of M.

Remark 8.8. Our notation deviates from Deligne's. Our $M^\natural$, as defined in Definition 8.7, corresponds to Deligne's $G^\natural$, and our $[L \to M^\natural]$ to his $M^\natural$. The reason for this is that we want to be able to distinguish between the universal vector extension $G^\natural$ of G (in $\mathcal{G}_\mathbb{Q}$) and $M^\natural$ when discussing $M = [L \to G]$.

Lemma 8.9. *The universal vector extension* $[L \to M^\natural]$ *of* $[L \to G]$ *is universal in the following sense: given a vector extension* $[L \to G']$, *there is a unique morphism*

$$[L \to M^\natural] \to [L \to G'].$$

Moreover, $L \to M^\natural$ *is injective.*

Proof The proof of the universal property here is the same as in the case of abelian varieties; see Proposition 4.20.

Let $l \in L$ be an element with image 0 in $M^\natural$. A fortiori, its image in G vanishes. Hence it suffices to show injectivity in the case $G = 0$. Then $\mathrm{Ext}^1(M, \mathbb{G}_a) = \mathrm{Hom}(L, \mathbb{G}_a)$, hence

$$M^\natural = \mathrm{Hom}(L, \mathbb{G}_a)^\vee.$$

The natural map $L \to M^\natural$ is the evaluation map $l \mapsto (\chi \mapsto \chi(l))$. It is injective.   □

The universal property of $G^\natural$ induces a canonical map $G^\natural \to M^\natural$. Comparing their vector group components, we see that we have a short exact sequence

$$0 \to G^\natural \to M^\natural \to \mathrm{Hom}(L, \mathbb{G}_a)^\vee \to 0.$$

By the structure theory of commutative algebraic groups, this even gives a (non-canonical) decomposition

$$M^\natural \cong G^\natural \times \mathrm{Hom}(L, \mathbb{G}_a)^\vee.$$

Lemma 8.10. *The universal vector extension defines faithful exact functors*

$$1\text{-Mot}_k \to \mathcal{G}, \quad M \mapsto M^\natural$$

and

$$1\text{-Mot}_k \to 1\text{-MOT}_k, \quad M = [L \to G] \mapsto [L \to M^\natural].$$

Proof The dimension of $M^\natural$ is given by the formula $\mathrm{rk}(L) + 2\dim(A^\natural) + \dim(T)$ for $M = [L \to G]$ with $0 \to T \to G \to A \to 0$ the decomposition into the torus part and the abelian part. If $M^\natural = 0$, then $M = 0$. This is faithfulness.

Let $0 \to M_1 \to M_2 \to M_3 \to 0$ be an exact sequence in 1-Mot_k. We have the commutative diagram

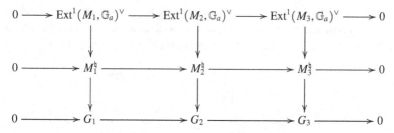

in which the sequence of semi-abelian varieties is the sequence of the semi-abelian parts of the motives. In the Ext-sequence the composition of the two maps vanishes by functoriality. In order to deduce exactness, it suffices to show that the dimensions add up in the right way. This is the case, because the dimension of $\mathrm{Ext}^1(M, \mathbb{G}_a)$ is linear in the dimension of the constituents. Together this implies exactness in the middle row. □

8.1.3 The de Rham Realisation

Definition 8.11. Let M be a 1-motive over k. We define the *de Rham realisation of M* as

$$V_{\mathrm{dR}}(M) := \mathrm{Lie}(M^\natural),$$

with $M^\natural$ as in Definition 8.7.

The de Rham realisation carries a weight filtration in the same way as the singular realistion, and in addition a *Hodge filtration*

$$F^p V_{\mathrm{dR}}(M) = \begin{cases} 0 & p > 0, \\ \ker(\mathrm{Lie}(M^\natural) \to \mathrm{Lie}(G)) & p = 0, \\ V_{\mathrm{dR}}(M) & p \leq -1; \end{cases}$$

see [Del74, Construction 10.1.8]. Note that $F^0 V_{\mathrm{dR}}(M) = \mathrm{Ext}^1_{1-\mathrm{MOT}}(M, \mathbb{G}_a)^\vee$.

Lemma 8.12. *The functor* $V_{\mathrm{dR}} \colon 1\text{-}\mathrm{Mot}_k \to k\text{-}\mathrm{Vect}$ *is faithful and exact.*

Proof Exactness follows from the exactness of the functors $M \mapsto M^\natural$ and $G \mapsto \mathrm{Lie}(G)$. Hence it suffices to check that $V_{\mathrm{dR}}(M) = 0$ implies $M = 0$ in $1\text{-}\mathrm{Mot}_k$. The assumption implies $M^\natural = 0$ and in consequence that G is trivial. In this case $M^\natural = \mathrm{Hom}(L, \mathbb{G}_a)^\vee$. Its vanishing also implies $L = 0$. □

8.1.4 The Period Isomorphism

In addition to the two realisations, there is a filtered comparison isomorphism $V_{\mathrm{sing}}(M)_{\mathbb{C}} \cong V_{\mathrm{dR}}(M)_{\mathbb{C}}$, the *period isomorphism*, which is constructed as follows: the structural map $L \to G$ has a canonical lift to the universal vector extension $M^\natural$; see Definition 8.7. We obtain the commutative diagram

$$
\begin{array}{ccc}
H_1^{\mathrm{sing}}(M^{\natural,\mathrm{an}}, \mathbb{Z}) & \xrightarrow{\;\cong\;} & H_1^{\mathrm{sing}}(G^{\mathrm{an}}, \mathbb{Z}) \\
\downarrow & & \downarrow \\
\mathrm{Lie}(M^\natural)_{\mathbb{C}} & \longrightarrow & \mathrm{Lie}(G)_{\mathbb{C}} \\
\downarrow{\scriptstyle \exp} & & \downarrow{\scriptstyle \exp} \\
L \longrightarrow M^{\natural,\mathrm{an}} & \longrightarrow & G^{\mathrm{an}}.
\end{array}
$$

The map at the top is an isomorphism by homotopy invariance because $M^\natural$ is a vector bundle over G. Hence the pull-back $T_{\mathrm{sing}}(M) = L \times_{G^{\mathrm{an}}} \mathrm{Lie}(G)_{\mathbb{C}}$ of $L \to G^{\mathrm{an}}$ to $\mathrm{Lie}(G)_{\mathbb{C}}$ agrees with the pull-back $L \times_{M^{\natural,\mathrm{an}}} \mathrm{Lie}(M^\natural)_{\mathbb{C}}$ of $L \to M^{\natural,\mathrm{an}}$ to $\mathrm{Lie}(M^\natural)_{\mathbb{C}}$. Let

$$\phi_M \colon V_{\mathrm{sing}}(M)_{\mathbb{C}} \to \mathrm{Lie}(M^\natural)_{\mathbb{C}}$$

be the map obtained by this identification of pull-backs.

Lemma 8.13 (Deligne [Del74, Construction 10.1.8]). *The morphism* ϕ_M *is a filtered isomorphism.*

Proof It is sufficient to consider the three cases $[L \to 0]$, $[0 \to T]$ for a torus T and $[0 \to A]$ for an abelian variety A separately. If they are isomorphisms, then ϕ_M is a filtered isomorphism in general.

For $M = [L \to 0]$, we have $V_{\text{sing}}(M) = L \otimes \mathbb{Q}$, $V_{\text{dR}}(M) = \text{Hom}(L, \mathbb{G}_a)^\vee$ and the period map is the natural map, hence an isomorphism after extension of scalars.

For $M = [0 \to T]$, we have $T_{\text{sing}}(M) = H_1(T^{\text{an}}, \mathbb{Z})$, $M^\natural = T$ and the period map is induced from the inclusion $H_1(T^{\text{an}}, \mathbb{Z}) \to \text{Lie}(T)^{\text{an}}$. It is an isomorphism after extension of scalars to $\mathbb{C}$.

We now turn to the case $M = [0 \to A]$. We have $V_{\text{dR}}(M) \otimes_k \mathbb{C} = \text{Lie}(A^\natural)^{\text{an}}$. The complex Lie group $(A^\natural)^{\text{an}}$ is the universal vector extension of A^{an} because $H^1(A_{\mathbb{C}}, \mathcal{O}) = H^1(A^{\text{an}}, \mathcal{O})$. To simplify notation we now write A instead of A^{an} and also use the abbreviations $T = T_{\text{sing}}(M) \cong H_1(A, \mathbb{Z})$ and $T_{\mathbb{C}} = T \otimes_{\mathbb{Z}} \mathbb{C}$. We want to show that

$$\phi: T_{\mathbb{C}} \to \text{Lie}(A^\natural)$$

is an isomorphism. Both sides have the same dimension, $2 \dim A$. The group $T \subset T_{\mathbb{C}}$ is in the kernel of the composition $T \to \text{Lie}(A^\natural) \xrightarrow{\exp_{A^\natural}} A$. We get an induced vector extension

$$T_{\mathbb{C}}/T \to A.$$

It suffices to show that it is the universal one. Let

$$0 \to V \to G \to A \to 0$$

be a vector extension. By homotopy invariance, the map $T \to \text{Lie}(A)$ lifts to $T \to \text{Lie}(G)$. (This is the same argument that we used to construct $\phi: T \to \text{Lie}(A^\natural)$ in the first place.) It induces a $\mathbb{C}$-linear map

$$T_{\mathbb{C}} \to \text{Lie}(G)$$

and a holomorphic group homomorphism

$$T_{\mathbb{C}}/T \to G.$$

In conclusion we have verified that $T_{\mathbb{C}}/T$ satisfies the universal property. $\square$

8.2 The Functor to Mixed Hodge Structures

Deligne introduced the category of 1-motives because of its close relation to Hodge theory. We use a modification that also takes the k-structure into account.

8.2.1 The Category of Mixed Hodge Structures

All filtrations on vector spaces are *separated* and *exhaustive*, i.e. they start with 0 and end with the full space. If $W_\bullet V$ is an ascending filtration, we write $\mathrm{Gr}_n^W V = W_n V / W_{n-1} V$. If $F^\bullet V$ is a descending filtration, we write $\mathrm{Gr}_F^p V = F^p V / F^{p+1} V$.

Definition 8.14 (Deligne [Del71]). Let $k \subset \mathbb{C}$ be a subfield. A *mixed $\mathbb{Q}$-Hodge structure defined over k* consists of

- a finite-dimensional $\mathbb{Q}$-vector space $V_{\mathbb{Q}}$ equipped with an ascending filtration $W_\bullet V_{\mathbb{Q}}$, the *weight filtration*,
- a finite-dimensional k-vector space V_{dR} equipped with an ascending filtration $W_\bullet V_{\mathrm{dR}}$ and a descending filtration $F^\bullet V_{\mathrm{dR}}$, the *Hodge filtration*,
- a filtered isomorphism $\phi \colon (V_{\mathbb{Q}}, W_\bullet) \otimes_{\mathbb{Q}} \mathbb{C} \to (V_{\mathrm{dR}}, W_\bullet) \otimes_k \mathbb{C}$,

such that for every $n \in \mathbb{Z}$ the data $(\mathrm{Gr}_n^W V_{\mathbb{Q}}, \mathrm{Gr}_n^W V_{\mathbb{C}}, \mathrm{Gr}_n^W \phi)$ with $V_{\mathbb{C}} = V_{\mathrm{dR}} \otimes \mathbb{C}$ and the induced Hodge filtration is a pure Hodge structure of weight n. This means that

$$\bigoplus_{p+q=n} (F^p \cap \bar{F}^q)\big(\mathrm{Gr}_n^W V_{\mathbb{C}}\big) = \mathrm{Gr}_n^W V_{\mathbb{C}}.$$

Here $\bar{F}^q$ is the complex conjugate of F^q with respect to the $\mathbb{R}$-structure induced by $\phi(V_{\mathbb{Q}})$.

A *morphism* $f \colon V \to V'$ of mixed $\mathbb{Q}$-Hodge structures over k consists of a filtered $\mathbb{Q}$-linear map $f_{\mathbb{Q}} \colon V_{\mathbb{Q}} \to V'_{\mathbb{Q}}$ and a bifiltered k-linear map $f_{\mathrm{dR}} \colon V_{\mathrm{dR}} \to V'_{\mathrm{dR}}$ compatible with the period isomorphisms of V and V'.

We denote the category of mixed $\mathbb{Q}$-Hodge structures over k by MHS_k.

The category is obviously additive and $\mathbb{Q}$-linear. Less obviously it is even abelian; see [Del71, Théorème 1.2.10].

Example 8.15. Let X be a smooth projective variety over $\mathbb{C}$. By the Hodge Decomposition Theorem, every cohomology class has a unique harmonic representative. The decomposition of harmonic forms into pq-forms gives

$$H_{\mathrm{sing}}^n(X^{\mathrm{an}}, \mathbb{C}) \cong \bigoplus_{p+q=n} H^{pq}$$

satisfying $\overline{H^{pq}} = H^{qp}$. With the Hodge filtration $F^p H_{\mathrm{sing}}^n(X^{\mathrm{an}}, \mathbb{C}) = \bigoplus_{p' \geq p} H^{p'q'}$ this turns $H_{\mathrm{sing}}^n(X^{\mathrm{an}}, \mathbb{Q})$ into a pure $\mathbb{Q}$-Hodge structure of weight n. Alternatively, the Hodge filtration is induced by the stupid filtration

$$F^p \Omega_X^* = [0 \to \cdots \to 0 \to \Omega_X^p \to \cdots]$$

on the complex of holomorphic or algebraic differential forms on X. If X is defined over a subfield of k, this point of view allows us also to define the Hodge filtration over the subfield.

The main result of [Del71] and [Del74] is the construction of a natural mixed Hodge structure on the cohomology of any algebraic variety over k.

8.2.2 The Relation with 1-Motives

By Lemma 8.13, the assignment $M \mapsto (V_{\text{sing}}(M), V_{\text{dR}}(M), \phi_M)$ defines a functor

$$V \colon 1\text{-Mot}_k \to \text{MHS}_k$$

from the category of iso-1-motives to the category of mixed Hodge structures over k.

Theorem 8.16 (Deligne [Del74, Construction 10.1.3, pp. 53–56]). *For $k = \mathbb{C}$, the functor $V \colon 1\text{-Mot}_\mathbb{C} \to \text{MHS}_\mathbb{C}$ is fully faithful. Its image consists of the full subcategory of the polarisable Hodge structures whose only non-zero Hodge numbers are $(-1, -1), (-1, 0), (0, -1), (0, 0)$.*

Note that a mixed Hodge structure with Hodge numbers as above is polarisable if and only if the graded piece in weight -1 is polarisable. Using the Analytic Subgroup Theorem, Deligne's result can be sharpened as follows.

Proposition 8.17. *In the case $k = \overline{\mathbb{Q}}$, the functor $V \colon 1\text{-Mot}_{\overline{\mathbb{Q}}} \to \text{MHS}_{\overline{\mathbb{Q}}}$ is fully faithful.*

Proof Let $M = [L \to G], M' = [L' \to G']$ be in $1\text{-Mot}_{\overline{\mathbb{Q}}}$ and

$$\gamma \colon V(M) \to V(M')$$

a morphism of Hodge structures over $\overline{\mathbb{Q}}$. By extension of scalars, we get a morphism of Hodge structures

$$(V_{\text{dR}}(M)_\mathbb{C}, V_{\text{sing}}(M), \phi) \xrightarrow{\gamma_\mathbb{C}} (V_{\text{dR}}(M')_\mathbb{C}, V_{\text{sing}}(M'), \phi)$$

over $\mathbb{C}$. By Deligne's theorem, $\gamma_\mathbb{C}$ is induced by a morphism γ_{Mot} in $1\text{-Mot}_\mathbb{C}$ and after replacing L by a rational multiple it is represented by

$$\gamma_{\text{Mot}} \colon [L \to G_\mathbb{C}] \to [L' \to G'_\mathbb{C}].$$

It remains to show that the induced morphism of algebraic groups $G_\mathbb{C} \to G'_\mathbb{C}$ is even defined over $\overline{\mathbb{Q}}$. Using the fact that

$$\text{Lie}(G) = V_{\text{dR}}(M)/F^0 V_{\text{dR}}(M),$$

the morphism of Hodge structures over $\overline{\mathbb{Q}}$ also induces a compatible homomorphism

$$\mathrm{Lie}(G) \to \mathrm{Lie}(G')$$

defined over $\overline{\mathbb{Q}}$. By the Analytic Subgroup Theorem (see Corollary 6.9), this is enough to imply that the group homomorphism is defined over $\overline{\mathbb{Q}}$. $\square$

By composition with the forgetful functor $\mathrm{MHS}_{\overline{\mathbb{Q}}} \to (\mathbb{Q},k)$-Vect (see Definition 7.1), we also get a functor

$$1\text{-}\mathrm{Mot}_{\overline{\mathbb{Q}}} \to (\mathbb{Q},k)\text{-Vect}.$$

We shall show in Theorem 9.14 that it is still fully faithful.

Remark 8.18. In the meantime André has shown in [And21] that $1\text{-}\mathrm{Mot}_k \to \mathrm{MHS}_{\mathbb{C}}$ is fully faithful even for all algebraically closed fields $k \subset \mathbb{C}$.

8.2.3 The Weight Filtration

The weight filtration on the Hodge structure is *motivic*, i.e. induced by a filtration on the motive.

Definition 8.19. Let $M = [L \to G]$ be a 1-motive, G an extension of the abelian variety A by a torus T, and define the *weight filtration of M* as

$$W_n M = \begin{cases} 0 & n \le -3, \\ [0 \to T] & n = -2, \\ [0 \to G] & n = -1, \\ M & n \ge 0. \end{cases}$$

The associated gradeds of this filtration are $[0 \to T]$, $[0 \to A]$ and $[L \to 0]$. The simple building blocks are $\mathbb{G}_m$, $[0 \to B]$ for simple abelian varieties B and $[\mathbb{Z} \to 0]$. The functors $W_n \colon 1\text{-}\mathrm{Mot}_k \to 1\text{-}\mathrm{Mot}_k$ are exact. This is often used to deduce the existence of splittings. For example:

Lemma 8.20. *Let $[L \to G]$ be a 1-motive and $[L' \to 0] \hookrightarrow M$ be an injective morphism. Then*

$$M \cong [L' \to 0] \times [L'' \to G],$$

where $L'' = L/L'$ (modulo torsion).

Proof We choose $L \cong L' \times L''$ (after replacing L by an isogenuous lattice). The natural map $[L'' \to G] \to M$ together with the given $[L' \to 0] \to M$ define $[L' \to 0] \times [L'' \to G] \to M$. This map is an isomorphism because it is an isomorphism on all gradeds with respect to the weight filtration. $\square$

8.3 The Key Comparison

We come back to the functor $(\cdot)^\natural \colon M \mapsto M^\natural$ from 1-motives to commutative algebraic groups. By Lemma 8.10, it is faithful and exact.

Proposition 8.21. *Let H be an object of $\mathcal{G}$ of the form $H = M^\natural$ for a 1-motive M. Given a short exact sequence*

$$0 \to H_1 \to H \to H_2 \to 0$$

in $\mathcal{G}$, there is a short exact sequence

$$0 \to M_1 \to M \to M_2 \to 0$$

in 1-Mot$_k$ and a commutative diagram

$$
\begin{array}{ccccccccc}
0 & \longrightarrow & H_1 & \longrightarrow & H & \longrightarrow & H_2 & \longrightarrow & 0 \\
& & \uparrow & & \| & & \uparrow & & \\
0 & \longrightarrow & M_1^\natural & \longrightarrow & M^\natural & \longrightarrow & M_2^\natural & \longrightarrow & 0
\end{array}
$$

such that

$$V_{\text{sing}}(M) \cap \text{Lie}(H_1)_{\mathbb{C}} = V_{\text{sing}}(M_1).$$

The sequence is uniquely determined by these properties.

Proof Let $M = [L \to G]$ be a 1-motive and

$$0 \to H_1 \to M^\natural \to H_2 \to 0$$

a short exact sequence of connected commutative algebraic groups. By the structure theory of commutative algebraic groups, there are canonical decompositions

$$0 \to V_i \to H_i \to G_i \to 0,$$

with G_i semi-abelian and V_i a vector group. Moreover, the sequence

$$0 \to G_1 \to G \to G_2 \to 0$$

is exact. The data organise as

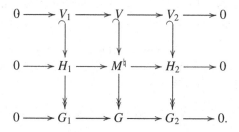

Let $L \to M^\natural$ be the canonical lift from G to $M^\natural$; see Definition 8.7. Note that it is injective, by Lemma 8.9. We define L_1 as the intersection of L with H_1 and L_2 as L/L_1 (modulo torsion). By construction, $L_1 \to L \to G$ factors via G_1 and $L \to G \to G_2$ via $L_2 \to G_2$. We put $M_i = [L_i \to G_i]$. By construction, the sequence

$$0 \to M_1 \to M \to M_2 \to 0$$

is exact and there are maps $L_i \to H_i$. By the universal property of $M_i^\natural$, this induces morphisms $M_i^\natural \to H_i$, compatible with the morphisms to/from $M^\natural$. For $i = 1$, the composition $M_1^\natural \to H_1 \to M^\natural$ is injective, hence so is $M_1^\natural \to H_1$. The dual argument gives surjectivity of $M_2^\natural \to H_2$.

We abbreviate $T_{\text{sing}}(H_1) = T_{\text{sing}}(M) \cap \text{Lie}(H_1)_{\mathbb{C}}$. Since $L_1 \to H_1^{\text{an}}$ is the pull-back of $L \to M^{\natural,\text{an}}$ and $\text{Lie}(H_1)_{\mathbb{C}} \to \text{Lie}(M^\natural)_{\mathbb{C}}$ is the pull-back of $H_1 \to M^\natural$ via the exponential map, we deduce that

$$T_{\text{sing}}(H_1) = T_{\text{sing}}(M) \cap \text{Lie}(H_1)_{\mathbb{C}} = \exp_{H_1}^{-1} L_1.$$

It follows that the sequence

$$0 \to \ker(\exp_{H_1}) \to T_{\text{sing}}(H_1) \to L_1 \to 0$$

is exact. We compare it with the same sequence for $M_1^\natural$. In both cases, the kernel computes $H_1^{\text{sing}}(G_1, \mathbb{Z})$ because they are vector groups over G_1. Hence they are the same, which implies the claim $T_{\text{sing}}(H_1) \cong T_{\text{sing}}(M_1)$ and hence $V_{\text{sing}}(M_1) \cong V_{\text{sing}}(M) \cap \text{Lie}(H_1)_{\mathbb{C}}$.

Given a second short exact sequence of motives

$$0 \to M_1' \to M \to M_2' \to 0$$

with the same properties, the assumptions imply $V_{\text{sing}}(M_1) = V_{\text{sing}}(M_1')$ inside $\text{Lie}(H_1)_{\mathbb{C}}$. Without loss of generality, even $T_{\text{sing}}(M_1) = T_{\text{sing}}(M_1')$. Let $M_1' = [L_1' \to G_1']$. We have $G_1' \subset G_1$ because this is the semi-abelian part of H_1. We also have $L_1' \subset L_1 = L \cap H_1$. This means $M_1' \subset M_1$. We get equality of iso-1-motives from equality of the singular realisations. $\square$

It is shown later (see Theorem 13.5) that the image of the functor 1-Mot$_{\overline{\mathbb{Q}}}$ $\to$ MHS$_{\overline{\mathbb{Q}}}$ is closed under subquotients.

Remarks 8.22. 1. The above result is the central input into our proof of the main result of our monograph: a version of the subgroup theorem for 1-motives in Theorem 9.7.

2. The proof even gives an integral construction of M_1 in a suitable abelian enlargement of the category of 1-motives such that

$$T_{\text{sing}}(M) \cap \text{Lie}(H_1)_{\mathbb{C}} = T_{\text{sing}}(M_1).$$

We do not need this for our applications to periods.

9

Periods of 1-Motives

In this chapter we introduce the set of periods of a 1-motive as obtained from the comparison isomorphism between the singular and the de Rham realisation of the motive. In Section 9.1 an alternative description of periods as explicit integrals is derived. This turns out to be more useful in transcendence theory. As a next step we determine in Section 9.2 the relations between periods. This is a classical open question which has been answered in the past in very special cases only. As a first main result of the chapter we show that all relations are induced by trivial ones. This gives an answer to Kontsevich's Period Conjecture in the case of 1-motives. For this we need an extension of the Subgroup Theorem to 1-motives. As a first application, we give in Section 9.3 a necessary and sufficient condition for periods of 1-motives to be algebraic.

9.1 Definition and First Properties

Let $M = [L \to G]$ be a 1-motive over $\overline{\mathbb{Q}}$.

Definition 9.1. The set of *periods of M* is the union of the sets of entries of the period matrices of the comparison isomorphism $V_{\text{sing}}(M)_{\mathbb{C}} \to V_{\text{dR}}(M)_{\mathbb{C}}$ with respect to all $\mathbb{Q}$-bases of $V_{\text{sing}}(M)$ and $\overline{\mathbb{Q}}$-bases of $V_{\text{dR}}(M)$. We denote it $\mathcal{P}(M)$ and denote by $\mathcal{P}\langle M \rangle$ the $\overline{\mathbb{Q}}$-subvector space of $\mathbb{C}$ generated by $\mathcal{P}(M)$.

For any subcategory $\mathcal{C} \subseteq 1\text{-Mot}_{\overline{\mathbb{Q}}}$, we write $\mathcal{P}(\mathcal{C})$ for the union of the $\mathcal{P}(M)$ for all objects $M \in C$, and $\mathcal{P}\langle\mathcal{C}\rangle$ for the vector space over $\overline{\mathbb{Q}}$ generated by $\mathcal{P}(\mathcal{C})$. In particular, we write $\mathcal{P}(1\text{-Mot}_{\overline{\mathbb{Q}}})$ for the union of all $\mathcal{P}(M)$.

Equivalently, we can define $\mathcal{P}(M)$ as the image of the period pairing

$$V_{\text{sing}}(M) \times V_{\text{dR}}^{\vee}(M) \to \mathbb{C}.$$

This description makes clear that it is not a vector space. We write $\omega(\sigma)$, or more suggestively $\int_\sigma \omega$, for the value of the period pairing for $\sigma \in V_{\text{sing}}(M)$ and $\omega \in V_{\text{dR}}^\vee(M)$. The notation will be justified later.

Note that this definition is an instance of the abstract definition of periods in Chapter 7.

Lemma 9.2. *The vector space* $\mathcal{P}\langle M \rangle$ *agrees with the set of periods of the additive subcategory generated by* M *and with the set of periods of the full abelian subcategory closed under subquotients generated by* M. *The set* $\mathcal{P}(1\text{-Mot}_{\overline{\mathbb{Q}}})$ *is a* $\overline{\mathbb{Q}}$-*subvector space of* $\mathbb{C}$.

Proof   Apply Lemma 7.5 to $\mathcal{C} = 1\text{-Mot}_{\overline{\mathbb{Q}}}$ and $V \colon 1\text{-Mot}_{\overline{\mathbb{Q}}} \to (\mathbb{Q}, k)\text{-Vect}$.   □

We want to make the definition of $\int_\sigma \omega$ explicit. Let $M = [L \to G]$ be a 1-motive over $\overline{\mathbb{Q}}$. Recall that $V_{\text{dR}}(M)$ was defined as $\text{Lie}(M^\natural)$ for a certain vector extension $M^\natural$ of G. Hence $V_{\text{dR}}^\vee(M) = \text{coLie}(M^\natural)$. Every cotangent vector of $M^\natural$ defines a unique $M^\natural$-equivariant differential form on $M^\natural$ and any equivariant global differential arises in this way. Therefore we may make the identification

$$V_{\text{dR}}^\vee(M) \cong \Omega^1(M^\natural)^{M^\natural}$$

with the space of invariant differentials and we may from now on view ω as a differential form on $M^\natural$. Recall also that $T_{\text{sing}}(M)$ is a lattice in $\text{Lie}(M^{\natural,\text{an}})$. The latter maps to $M^{\natural,\text{an}}$ via the exponential map for the Lie group $M^{\natural,\text{an}}$; see Section 5.2. Hence a vector field $u \in T_{\text{sing}}(M)$ defines a point

$$\exp(u) \in M^{\natural,\text{an}}$$

and, by composition of a straight path from 0 to u with the exponential map, a path

$$\gamma_u \colon [0, 1] \to M^{\natural,\text{an}}$$

from 0 to $\exp(u)$. The period pairing is then computed as

$$\int_u \omega = \int_{\gamma_u} \omega,$$

where the right-hand side is an honest integral on a manifold; see Section 5.3 for more details. In particular, for $u \in \ker(\exp \colon \text{Lie}(M^{\natural,\text{an}}) \to M^{\natural,\text{an}})$, the path γ_u is closed and hence it defines an element of $H_1(M^{\natural,\text{an}}, \mathbb{Z}) \cong H_1(G^{\text{an}}, \mathbb{Z})$. Conversely, every element $\sigma \in H_1(M^{\natural,\text{an}}, \mathbb{Z})$ is represented by a formal linear combination $\sigma = \sum a_i \gamma_i$ of closed loops γ_i. The cycle defines a linear map

$$I(\sigma) \colon \text{coLie}(M^{\natural,\text{an}}) \to \mathbb{C},$$

$$\omega \mapsto \sum a_i \int_{\gamma_i} \omega.$$

In other words, $I(\sigma) \in \mathrm{Lie}(M^{\natural,\mathrm{an}})^{\vee\vee} \cong \mathrm{Lie}(M^{\natural,\mathrm{an}})$. As spelt out in Section 5.3, the two operations are inverse to each other: $I(\gamma_u) = u$ for $u \in \ker(\exp)$ and $\gamma(I(\sigma))$ is homologous to $\sigma \in H_1(M^{\natural,\mathrm{an}}, \mathbb{Z})$.

Lemma 9.3. *We have* $\exp(u) \in M^{\natural}(\overline{\mathbb{Q}})$.

Proof   By definition, we have the commutative diagram

$$
\begin{array}{ccc}
T_{\mathrm{sing}}(M) & \longrightarrow & \mathrm{Lie}(M^{\natural,\mathrm{an}}) \\
\downarrow & & \downarrow{\scriptstyle\exp} \\
L & \longrightarrow & M^{\natural,\mathrm{an}},
\end{array}
$$

which we evaluated at $u \in T_{\mathrm{sing}}(M)$. The value is in $M^{\natural}(\overline{\mathbb{Q}})$ because L takes values there.   □

For the record, the above argument proves the following result.

Proposition 9.4. *Every period of a* 1-*motive is of the form*

$$
\int_{\gamma} \omega,
$$

where ω *is an algebraic* 1-*form on a commutative algebraic group* G *over* $\overline{\mathbb{Q}}$ *and* γ *is a path from* 0 *to a point* $P \in G(\overline{\mathbb{Q}})$.

Example 9.5. Let M be $[\mathbb{Z} \to 0]$. Then $M^{\natural} = \mathbb{G}_a$, $V_{\mathrm{sing}}(M) = \mathbb{Q}$, $V_{\mathrm{dR}}^{\vee}(M) = \mathrm{coLie}(\mathbb{G}_a) = \overline{\mathbb{Q}}dt$ and the period map sends 1 to $\partial/\partial t$. The image of the period pairing $V_{\mathrm{sing}}(M) \times V_{\mathrm{dR}}^{\vee}(M) \to \mathbb{C}$ is simply $\overline{\mathbb{Q}}$. In terms of integration: an element $u \in T_{\mathrm{sing}}(M) \subset \mathrm{Lie}(\mathbb{G}_a)_{\mathbb{C}}$ gives rise to a path γ_u from 0 to u in $\mathbb{C}$. An element $\omega \in V_{\mathrm{dR}}^{\vee}(M)$ is identified with a differential form αdt on $\mathbb{G}_a$, and the period is

$$
\int_{0}^{u} \alpha dt = \alpha u \in \overline{\mathbb{Q}}.
$$

More generally, all periods of motives of the form $[L \to 0]$ are algebraic.

We are going to see many more explicit examples in subsequent chapters.

9.2 Relations Between Periods

There are two types of obvious sources of relations between periods of 1-motives:

1. (Bilinearity) Let M be a 1-motive over $\overline{\mathbb{Q}}$, $\sigma_1, \sigma_2 \in V_{\text{sing}}(M)$, and take $\omega_1, \omega_2 \in V_{\text{dR}}^{\vee}(M)$, $\mu_1, \mu_2 \in \mathbb{Q}$, $\lambda_1, \lambda_2 \in \overline{\mathbb{Q}}$. Then

$$\int_{\mu_1 \sigma_1 + \mu_2 \sigma_2} (\lambda_1 \omega_1 + \lambda_2 \omega_2) = \sum_{i,j=1,2} \mu_i \lambda_j \int_{\sigma_i} \omega_j.$$

2. (Functoriality) Let $f: M \to M'$ be a morphism in 1-Mot$_{\overline{\mathbb{Q}}}$. Let $\sigma \in V_{\text{sing}}(M)$ and $\omega' \in V_{\text{dR}}^{\vee}(M)$. Then

$$\int_{\sigma} f^* \omega' = \int_{f_* \sigma} \omega'.$$

As a special case we get the relations coming from short exact sequences: consider

$$0 \to M' \overset{i}{\to} M \overset{p}{\to} M'' \to 0$$

in 1-Mot$_k$. Then the period matrix for M will be block triangular, hence will contain plenty of zeroes. Explicitly: for $\sigma' \in V_{\text{sing}}(M')$ and $\omega'' \in V_{\text{dR}}^{\vee}(M'')$, we have

$$\int_{i_* \sigma'} p^* \omega'' = \int_{\sigma'} i^* p^* \omega'' = \int_{p_* i_* \sigma'} \omega'' = 0$$

because $i^* p^* \omega'' = 0$ and $p_* i_* \sigma' = 0$.

This is actually the *only* source of relations between periods of 1-motives, as we will show.

Definition 9.6. For $u \in V_{\text{sing}}(M)$ we define $\text{Ann}(u) \subseteq V_{\text{dR}}^{\vee}(M)$ as the left kernel of the period pairing.

Theorem 9.7 (Subgroup Theorem for 1-motives). *Given a 1-motive M over $\overline{\mathbb{Q}}$ and $u \in V_{\text{sing}}(M)$, there exists an exact sequence in 1-Mot$_{\overline{\mathbb{Q}}}$*

$$0 \to M_1 \overset{i}{\to} M \overset{p}{\to} M_2 \to 0,$$

such that $\text{Ann}(u) = p^* V_{\text{dR}}^{\vee}(M_2)$ *and* $u \in i_* V_{\text{sing}}(M_1)$. *It is uniquely determined by these properties.*

Proof We consider the connected commutative algebraic group $M^{\natural}$. Without loss of generality, $u \in T_{\text{sing}}(M)$. By Lemma 9.3, $u \in \text{Lie}(M^{\natural})_{\mathbb{C}}$ with $\exp(u) \in M^{\natural}(\overline{\mathbb{Q}})$. We apply the Analytic Subgroup Theorem in the version of Theorem 6.2. Hence there is a short exact sequence in $\mathcal{G}$,

$$0 \to H_1 \to M^{\natural} \overset{\pi}{\to} H_2 \to 0,$$

such that $u \in \text{Lie}(H_1)_{\mathbb{C}}$ and $\text{Ann}(u) = \pi^* \text{coLie}(H_2)$. By Proposition 8.21, we find an associated short sequence in 1-Mot$_{\overline{\mathbb{Q}}}$,

$$0 \to M_1 \to M^{\natural} \to M_2 \to 0,$$

such that $T_{\text{sing}}(M_1) = T_{\text{sing}}(M) \cap \text{Lie}(H_1)_{\mathbb{C}}$. In particular, $u \in T_{\text{sing}}(M_1)$. Because of the short exact sequence of motives, we have $\text{coLie}(M_2^\natural) = V_{\text{dR}}^\vee(M_2) \subset \text{Ann}(u)$. On the other hand, Proposition 8.21 also gives a surjection $M_2^\natural \to H_2$, hence

$$\text{Ann}(u) = \text{coLie}(H_2) \subset \text{coLie}(M_2^\natural) \subset \text{Ann}(u).$$

This implies equality of co-Lie algebras and hence even $M_2^\natural \cong H_2$.

Suppose that there is a second short exact sequence

$$0 \to M_1' \xrightarrow{i'} M \xrightarrow{p'} M_2' \to 0$$

with the same properties. In particular, $p'^* V_{\text{dR}}^\vee(M_2') = p^* V_{\text{dR}}^\vee(M_2)$ inside $V_{\text{dR}}^\vee(M)$. This means that $M_2^\natural = M_2'^\natural$ as quotients of $M^\natural$. As the functor $\cdot^\natural$ is faithful, this gives $M_2 = M_2'$ as quotients of M. $\quad\square$

Remark 9.8. The proof shows that the decompositions in terms of algebraic groups (Theorem 6.2) and in terms of 1-motives agree.

Most of the time we apply the theorem through the following consequence.

Corollary 9.9. *Given a* 1-*motive* M *over* $\overline{\mathbb{Q}}$, $u \in V_{\text{sing}}(M)$, $\omega \in V_{\text{dR}}^\vee(M)$ *such that* $\int_u \omega = 0$, *there is a short exact sequence*

$$0 \to M_1 \xrightarrow{i} M \xrightarrow{p} M_2 \to 0$$

of 1-*motives and* $u_1 \in V_{\text{sing}}(M_1)$, $\omega_2 \in V_{\text{dR}}^\vee(M_2)$ *such that* $u = i_* u_1$, $\omega = p^* \omega_2$.

Proof We apply the theorem. This already gives the existence of u_1. By assumption, we have $\omega \in \text{Ann}(u)$, hence ω is in the image of p^*. $\quad\square$

Our main aim is to prove the following, which was formulated as a conjecture in [Wüs12]. We remind the reader that $\langle M \rangle$ is the full subcategory of 1-Mot$_{\overline{\mathbb{Q}}}$ generated by M and closed under subquotients.

Theorem 9.10 (Kontsevich's Period Conjecture for 1-motives)**.** *All* $\overline{\mathbb{Q}}$-*linear relations between elements of* $\mathcal{P}(1\text{-Mot}_{\overline{\mathbb{Q}}})$ *are induced by bilinearity and functoriality.*

More precisely, for every 1-*motive* M *the relations between elements of* $\mathcal{P}(M)$ *are generated by bilinearity and functoriality for morphisms in* $\langle M \rangle$, *or, equivalently, morphisms of the induced mixed Hodge structures over* $\overline{\mathbb{Q}}$. *In other words, the Period Conjecture in the sense of Definition 7.15 holds for* $\langle M \rangle$.

Proof We consider a linear relation between periods. For $i = 1, \ldots, n$ let $\alpha_i = \int_{\sigma_i} \omega_i$ be periods for 1-motives M_i and let $\lambda_i \in \overline{\mathbb{Q}}$ be such that

$$\lambda_1 \alpha_1 + \lambda_2 \alpha_2 + \cdots + \lambda_n \alpha_n = 0. \tag{9.1}$$

We have already argued in Lemma 7.5 that a linear combination of periods can be represented as single period. We now have to go through the construction carefully in order to check that no relations other than bilinearity and functoriality are used.

We put $M = M_1 \oplus \cdots \oplus M_n$. By pull-back via the projection, we can view each ω_i as an element of $V_{\mathrm{dR}}^\vee(M)$. By push-forward via the inclusion we may view each σ_i in $V_{\mathrm{sing}}(M)$. We then put $\sigma = \sum \sigma_i$ and $\omega = \sum \lambda_j \omega_j$. From the additivity relation we deduce

$$\int_\sigma \omega = \sum_{i,j} \lambda_j \int_{\sigma_i} \omega_j.$$

The functoriality relation leads to $\int_{\sigma_i} \omega_j = 0$ for $i \neq j$ and gives α_i for $i = j$. Hence the left-hand side of (9.1) equals $\int_\sigma \omega$.

We are now in the situation

$$\int_\sigma \omega = 0$$

on the 1-motive M. In other words, $\omega \in \mathrm{Ann}(\sigma)$. By Theorem 9.7, there is a short exact sequence

$$0 \to M' \xrightarrow{i} M \xrightarrow{p} M'' \to 0$$

in 1-Mot$_{\overline{\mathbb{Q}}}$ such that $\sigma = i_* \sigma'$ for $\sigma' \in V_{\mathrm{sing}}(M')$ and $\omega = p^* \omega''$ for $\omega'' \in V_{\mathrm{dR}}^\vee(M'') = \mathrm{Ann}(\sigma)$. Hence the vanishing of

$$\int_\sigma \omega = \int_{i_* \sigma'} p^* \omega'' = \int_{\sigma'} 0$$

is now implied by functoriality of 1-motives. $\qquad\qquad\qquad\qquad \square$

9.3 Transcendence of Periods of 1-Motives

All this implies a result on transcendence of periods.

Theorem 9.11 (Transcendence). *Let* $M = [L \to G]$ *be a* 1-*motive,* $\sigma \in V_{\mathrm{sing}}(M)$ *and* $\omega \in V_{\mathrm{dR}}^\vee(M)$. *Then the integral*

$$\int_\sigma \omega$$

is in $\overline{\mathbb{Q}}$ *if and only if there are* $\phi, \psi \in V_{\mathrm{dR}}^\vee(M)$ *with*

$$\omega = \phi + \psi$$

such that $\int_\sigma \psi = 0$ *and the image of* ϕ *in* $V_{\mathrm{dR}}^\vee(G)$ *vanishes.*

Proof We write $\alpha = \int_\sigma \omega$ and we begin with the easy direction. The short exact sequence

$$0 \to [0 \to G] \to M \to [L \to 0] \to 0$$

induces a short exact sequence

$$0 \to V_{dR}^\vee([L \to 0]) \to V_{dR}^\vee(M) \to V_{dR}^\vee([0 \to G]) \to 0.$$

Suppose that the image of ϕ in $V_{dR}^\vee([0 \to G])$ vanishes or, equivalently, that $\phi \in V_{dR}^\vee([L \to 0])$.

Let $\bar{\sigma}$ be the image of σ in $V_{sing}([L \to 0])$. Then

$$\alpha = \int_\sigma \phi = \int_{\bar{\sigma}} \phi$$

is a period for $[L \to 0]$. It is a general fact; see Example 9.5, that $V_{dR}^\vee([\mathbb{Z}^r \to 0]) = \text{coLie}(\mathbb{A}^r)$ and that all its periods are algebraic.

Conversely, assume that α is algebraic. If $\alpha = 0$, the theorem holds with $\omega = \psi$. Assume that it is non-zero from now on. The algebraicity of α means that we can write α as $\int_{\sigma'} \omega'$ with $\omega' = \alpha dt$ and $\sigma' \in V_{sing}([\mathbb{Z} \to 0])$ the standard basis vector. Let $\Sigma = (\sigma, -\sigma') \in V_{sing}(M \oplus [\mathbb{Z} \to 0])$ and $\Omega = (\omega, \omega') \in V_{dR}^\vee(M \oplus [\mathbb{Z} \to 0])$. By assumption,

$$\int_\Sigma \Omega = \int_\sigma \omega - \int_{\sigma'} \omega' = 0.$$

By the Subgroup Theorem for 1-motives, Theorem 9.7, there is a short exact sequence of iso-1-motives

$$0 \to M_1 \xrightarrow{(\iota, \iota_\mathbb{Z})} M \oplus [\mathbb{Z} \to 0] \xrightarrow{p+q} M_2 \to 0, \tag{9.2}$$

and $\sigma_1 \in V_{sing}(M_1)$, $\omega_2 \in V_{dR}^\vee(M_2)$ such that

$$(\iota, \iota_\mathbb{Z})_* \sigma_1 = (\sigma, -\sigma'), \quad p^*\omega_2 = \omega, \quad q^*\omega_2 = \omega'.$$

The map $q \colon [\mathbb{Z} \to 0] \to M_2$ does not vanish because the pull-back of ω_2 is ω'. The latter is non-zero because α is assumed to be non-zero. The non-vanishing of the map already implies that $[\mathbb{Z} \to 0]$ is a direct summand of M_2. We explain the argument as follows. We write M_2 in the form $[L_2 \to G_2]$. Then the composition

$$[\mathbb{Z} \to 0] \to [L_2 \to G_2] \to [L_2 \to 0]$$

is non-trivial. The composition $\mathbb{Z} \to L_2$ does not vanish, hence we can decompose L_2 up to isogeny into the image of $\mathbb{Z}$ and a direct complement. This defines a section of our map.

We decompose M_2 as $[\mathbb{Z} \to 0] \oplus M_2'$. Hence $\omega_2 = \phi_2 + \psi_2$, with ϕ_2 coming from $[\mathbb{Z} \to 0]$ and ψ_2 from the complement M_2'. Let $\phi = p^*\phi_2$ and $\psi = p^*\psi_2$ be their images in $V_{dR}^\vee(M)$. Then $\omega = p^*\omega_2 = \phi + \psi$ and hence

$$\alpha = \int_\sigma \omega = \int_\sigma \phi + \int_\sigma \psi.$$

By splitting off the direct summand $[\mathbb{Z} \to 0]$ from the sequence (9.2), we obtain the short exact sequence

$$0 \to M_1 \to M \to M_2' \to 0.$$

The element $\sigma = \iota_*\sigma_1 \in V_{sing}(M)$ is induced from M_1, and $\psi = p^*\psi_2$ from M_2' and we conclude that $\int_\sigma \psi = 0$.

The element $\phi = p^*\phi_2$ is induced from $[\mathbb{Z} \to 0]$, hence it is in the image of

$$V_{dR}^\vee([\mathbb{Z} \to 0]) \to V_{dR}^\vee([L \to 0]) \to V_{dR}^\vee(M).$$

It follows that its image in $V_{dR}^\vee(G)$ vanishes. □

The general period formalism explained Chapter 7 also implies a dimension formula. We put (see Definition 7.23 with $T = V_{sing}|_{\langle M \rangle}$)

$$E(M) = \left\{ (\phi_N) \in \prod_{N \in \langle M \rangle} \text{End}_\mathbb{Q}(V_{sing}(N)) \,|\, \phi_{N'} \circ f = f \circ \phi_N \, \forall f \colon N \to N' \right\}.$$

This is a subalgebra of $\text{End}_\mathbb{Q}(V_{sing}(M))$, hence finite dimensional over $\mathbb{Q}$. In fact it agrees with $\text{End}(V_\mathbb{Q}|_{\langle M \rangle})$ in the notation of Section 7.2.

By Theorem 7.29, the category $\langle M \rangle$ is equivalent to the category of finitely generated $E(M)$-modules.

Corollary 9.12. *We have*

$$\dim_{\overline{\mathbb{Q}}} \mathcal{P}\langle M \rangle = \dim_\mathbb{Q} E(M).$$

Proof By Theorem 9.10 the Period Conjecture holds for $\langle M \rangle$. We now apply the abstract dimension formula of Proposition 7.31. Note that $E(M) = \text{End}(V_\mathbb{Q}|_{\langle M \rangle})$. □

Remark 9.13. This is a clear qualitative characterisation. However, it is by no means obvious how to compute the explicit value for a given M. We carry out this computation in Part III.

9.4 Fullness

As pointed out in Corollary 7.19, the validity of the Period Conjecture for a category $\mathcal{C}$ implies fullness of the functor $F \colon \mathcal{C} \to (K, L)$-Vect under consideration. In our case, we give a direct proof from the Analytic Subgroup Theorem for 1-motives.

Theorem 9.14 (Fullness). *The functor* $V: 1\text{-Mot}_{\overline{\mathbb{Q}}} \to (\mathbb{Q}, \overline{\mathbb{Q}})\text{-Vect}$ *is fully faithful with image closed under subquotients.*

Proof Note that the functor V is faithful and exact because the functors V_{sing} and V_{dR} are.

We choose $M \in 1\text{-Mot}_{\overline{\mathbb{Q}}}$ and consider a short exact sequence in $(\mathbb{Q}, \overline{\mathbb{Q}})\text{-Vect}$,

$$0 \to V' \xrightarrow{\iota} V(M) \xrightarrow{\pi} V'' \to 0.$$

It gives a two-step filtration of $V(M)$, which furnishes a period matrix for $V(M)$ in block-triangular form, i.e. the period matrix contains a square of zeroes. We show that $V' = V(M')$ and $V'' = V(M'')$ for objects M', M'' of 1-Mot_k. To see this, consider $\text{Ann} = \text{Ann}(\iota_* V'_{\mathbb{Q}}) \subset V^\vee_{\text{dR}}(M)$. We record for later use that

$$\pi^*(V''^\vee_{\overline{\mathbb{Q}}}) \subset \text{Ann}. \tag{9.3}$$

By Theorem 9.7 applied to the elements of $V'_{\mathbb{Q}}$, there is a short exact sequence

$$0 \to M' \xrightarrow{i} M \xrightarrow{p} M'' \to 0$$

in 1-$\text{Mot}_{\overline{\mathbb{Q}}}$ such that $p^* V^\vee_{\text{dR}}(M'') = \text{Ann}$ and $V'_{\mathbb{Q}} \subset V_{\text{sing}}(M')$. We have

$$\iota_* V'_{\mathbb{Q}} \subset \ker(V_{\text{sing}}(M) \to V_{\text{sing}}(M'')) = i_* V_{\text{sing}}(M').$$

Note that both $V(M')$ and V' are subobjects of $V(M)$. We apply Lemma 2.13 to the faithful exact functor $V \mapsto V_{\mathbb{Q}}$. This gives even $V' \subset V(M')$. This also implies $V(M'') \twoheadrightarrow V''$ and hence

$$\text{Ann} = p^* V^\vee_{\text{dR}}(M'') \subset \pi^*(V''^\vee_{\overline{\mathbb{Q}}}).$$

Together with equation (9.3) this gives $p^* V^\vee_{\text{dR}}(M'') = \pi^*(V''^\vee_{\overline{\mathbb{Q}}})$ inside $V^\vee_{\text{dR}}(M)$. Applying Lemma 2.13 again, this time to the faithful exact functor $V \mapsto V^\vee_{\overline{\mathbb{Q}}}$, this even leads to $V(M'') = V''$ and in turn $V(M') = V'$. We have now shown that the image of f_3 is closed under subobjects and quotients, implying the same for subquotients.

Closedness of the image under subquotients implies fullness, by Lemma 2.14.   □

Remark 9.15. Theorem 9.14 is not equivalent to Theorem 9.10. Consider, for example, the full abelian subcategory $\mathcal{A}$ closed under subquotients of

$(\mathbb{Q}, \overline{\mathbb{Q}})$-Vect generated by the single object $V = (\overline{\mathbb{Q}}^2, \mathbb{Q}^2, \phi)$, with $\phi \colon \mathbb{C}^2 \to \mathbb{C}^2$ given by multiplication by the matrix

$$\Phi = \begin{pmatrix} 1 & \pi \\ \log 2 & 1 \end{pmatrix}.$$

It is easy to check that V is simple and that its only endomorphisms are multiplication by rational numbers. Indeed, a subobject $V' \subsetneq V$ is determined by a vector $v = \begin{pmatrix} s \\ t \end{pmatrix} \in \overline{\mathbb{Q}}^2$ such that

$$\phi(v) = \begin{pmatrix} 1 & \pi \\ \log 2 & 1 \end{pmatrix} \begin{pmatrix} s \\ t \end{pmatrix} = \begin{pmatrix} s + t\pi \\ s \log 2 + t \end{pmatrix} \in \mathbb{Q}^2.$$

As π and $\log 2$ are transcendental, this implies $t = s = 0$. An endomorphism $f \colon V \to V$ is represented by a pair of matrices $A_{\overline{\mathbb{Q}}} = \begin{pmatrix} \alpha & \beta \\ \gamma & \delta \end{pmatrix}$ on $V_{\overline{\mathbb{Q}}}$ and $A_{\mathbb{Q}} = \begin{pmatrix} a & b \\ c & d \end{pmatrix}$ on $V_{\mathbb{Q}}$ such that $\Phi A_{\overline{\mathbb{Q}}} = A_{\mathbb{Q}} \Phi$.

Spelling out the matrix equation we get

$$\begin{pmatrix} \alpha + \pi\gamma & \beta + \pi\delta \\ \log 2\alpha + \gamma & \log 2\beta + \delta \end{pmatrix} = \begin{pmatrix} a + b \log 2 & a\pi + b \\ c + d \log 2 & c\pi + d \end{pmatrix}.$$

By the $\overline{\mathbb{Q}}$-linear independence of $1, \pi$ and $\log 2$ it implies that

$$\gamma = b = 0, \quad \alpha = a, \quad \delta = a, \quad \beta = b, \quad \alpha = d, \quad \gamma = c, \quad \beta = c = 0, \quad d = \delta$$

and shows that $A_{\overline{\mathbb{Q}}} = A_{\mathbb{Q}}$ is a diagonal matrix for $a \in \mathbb{Q}$. In other words, the full abelian category $\langle V \rangle \subset (\mathbb{Q}, \overline{\mathbb{Q}})$-Vect generated by V is semi-simple with a single simple object V for which $\mathrm{End}(V) = \mathbb{Q}$.

The space of periods $\mathcal{P}\langle V \rangle$ of V has $\overline{\mathbb{Q}}$-dimension 3 with basis $1, \pi, \log 2$. We compare it with the space $\widetilde{\mathcal{P}}(\langle V \rangle)$, which was introduced in Definition 7.6. The relations listed above imply that $\widetilde{\mathcal{P}}(\langle V \rangle)$ is generated by four elements corresponding to the four entries of the period matrix of V. There are no additional relations coming from subobjects or endomorphisms. This proves that $\dim_{\overline{\mathbb{Q}}} \widetilde{\mathcal{P}}(\langle V \rangle) = 4$, which is not the same as $3 = \dim_{\overline{\mathbb{Q}}} \mathcal{P}\langle V \rangle$, hence the Period Conjecture does not hold for $\langle V \rangle$. As it does hold for 1-Mot$_{\overline{\mathbb{Q}}}$, this implies that Φ does not occur as the period matrix of a 1-motive, even though all entries are indeed in $\mathcal{P}^1$.

Remark 9.16. The fact that the Period Conjecture implies Theorem 9.14 is a special case of a general pattern; see Corollary 7.19. For a careful analysis of this property we refer the reader to the discussion in [HMS17, Proposition 13.2.8] and [Hub20]. The relation between the Period Conjecture and the Hodge conjecture is also explained there. The fullness question is also taken up by Andreatta, Barbieri-Viale and Bertapelle in their recent work [ABVB20]. They give an independent proof of Theorem 13.5. Their second proof copies ours, but without the detour to periods.

10

First Examples

Before turning to the case of period numbers of curves more generally, we give some examples, all of them very classical. They do not rely on the full strength of Theorem 9.7, but could be deduced directly from the Analytic Subgroup Theorem as in Theorem 6.2. We still prefer to go via Theorem 9.7 in order to demonstrate the method.

10.1 Squaring the Circle

To prove the transcendence of π, or rather more naturally in our setting of $2\pi i$, we take the 1-motive $M_1 = [0 \to \mathbb{G}_m]$. Then $M_1^\natural = \mathbb{G}_m$ because $\operatorname{Ext}^1_{\mathcal{G}}(\mathbb{G}_m, \mathbb{G}_a) = 0$ and this means that by definition we have

$$V_{\mathrm{dR}}^\vee(M_1) = \operatorname{coLie}(\mathbb{G}_m) = \Omega^1(\mathbb{G}_m)^{\mathbb{G}_m}$$

for the de Rham realisation of the motive. On the singular side we get

$$V_{\mathrm{sing}}(M_1) = \ker(\operatorname{Lie}(\mathbb{G}_m)^{\mathrm{an}} \to \mathbb{G}_m^{\mathrm{an}}) = \ker(\exp \colon \mathbb{C} \to \mathbb{C}^*).$$

We take $\omega_1 = dz/z \in V_{\mathrm{dR}}^\vee(M_1)$ and the positively oriented loop $\gamma_1 \colon [0,1] \to \mathbb{C}^*$ given by $\gamma_1(s) = e^{2\pi i s}$ around 0. In the notation of Section 5.3, we define $\sigma_1 = I(\gamma_1) \in \operatorname{Lie}(\mathbb{G}_m)^{\mathrm{an}}$ and obtain

$$\int_{\sigma_1} \omega_1 = \int_{\gamma_1} \frac{dz}{z} = 2\pi i.$$

Corollary 10.1 (Lindemann 1882). *The period $2\pi i$ is transcendental.*

First proof. Assume that $2\pi i$ is algebraic. Then, by Theorem 9.11, we can express ω_1 as

$$\omega_1 = \phi + \psi \in V_{\mathrm{dR}}^\vee(M_1),$$

93

such that the image of ϕ vanishes in $V_{\text{dR}}^{\vee}(\mathbb{G}_m)$ and such that $\int_{\sigma_1} \psi = 0$. As $\mathbb{G}_m = M_1^{\natural}$, the map $V_{\text{dR}}^{\vee}(M_1) \to V_{\text{dR}}^{\vee}(\mathbb{G}_m)$ is an isomorphism. Hence $\phi = 0$. This means that $\omega_1 = \psi$ and hence $2\pi i = 0$, which is false. $\qquad\square$

We offer a second proof where the result is deduced from the Analytic Subgroup Theorem for 1-motives. Let $M_2 = [\mathbb{Z} \to 0]$. Then $M_2^{\natural} = \mathbb{G}_a$ because $\text{Ext}_{1-\text{MOT}}^1(M_2, \mathbb{G}_a) = \text{Hom}(\mathbb{Z}, \mathbb{G}_a) = \mathbb{G}_a$. This gives

$$V_{\text{dR}}^{\vee}(M_2) = \text{coLie}(\mathbb{G}_a) = \Omega^1(\mathbb{G}_a)^{\mathbb{G}_a}.$$

In this case $\exp \colon \text{Lie}(M_2^{\natural})^{\text{an}} \to M_2^{\natural,\text{an}}$ is the identity on $\mathbb{G}_a^{\text{an}} = \mathbb{C}$. The lattice $\mathbb{Z}$ embeds naturally into $M_2^{\natural,\text{an}} = \mathbb{C}$. In conclusion,

$$V_{\text{sing}}(M_2) = \exp^{-1}(\mathbb{Z}) = \mathbb{Z}.$$

Let γ_2 be the straight path from 0 to 1 in $\mathbb{G}_a^{\text{an}} = \mathbb{C}$ with image $\sigma_2 = I(\gamma_2) \in \mathfrak{g}_a^{\text{an}}$. For ω_2 we take $dt \in V_{\text{dR}}^{\vee}(M_2)$ to get

$$\int_{\sigma_2} \omega_2 = \int_{\gamma_2} dt = 1.$$

Second proof of Corollary 10.1. Assume that $2\pi i$ is algebraic. This means that

$$2\pi i + \alpha = 0$$

for some $\alpha \in \overline{\mathbb{Q}}$. In the notation from above, we consider

$$M = M_1 \times M_2 = [\mathbb{Z} \xrightarrow{0} \mathbb{G}_m]$$

and deduce

$$V_{\text{dR}}^{\vee}(M) = V_{\text{dR}}^{\vee}(M_1) \times V_{\text{dR}}^{\vee}(M_2), \quad V_{\text{sing}}(M) = V_{\text{sing}}(M_1) \times V_{\text{sing}}(M_2).$$

We choose

$$\sigma = (\sigma_1, \sigma_2) \in V_{\text{sing}}(M)$$

and

$$\omega = (\omega_1, \alpha\omega_2) \in V_{\text{dR}}^{\vee}(M).$$

Here we have used that α is algebraic. We find that

$$\int_{\sigma} \omega = \int_{\sigma_1} \omega_1 + \int_{\sigma_2} \alpha\omega_2 = 2\pi i + \alpha \cdot 1 = 0.$$

Now Theorem 9.7 is applied to the motive M and the classes ω and σ. Their period vanishes, hence there is a short exact sequence

$$0 \to M' \xrightarrow{i} M \xrightarrow{p} M'' \to 0$$

such that σ is in the image of i_* and ω is in the image of p^*.

Note that M is the product of two simple non-isomorphic 1-motives, hence this leaves only four possible choices for M':

$$0, \quad M_1 \times 0, \quad 0 \times M_2, \quad M.$$

We go through the cases. If $M' = 0$, then the image of i_* is zero and hence $\sigma = 0$, in contradiction to our hypothesis.

If $M' = M_1 \times 0$, then the second component σ_2 of σ is zero. This is false. The same argument also eliminates $M' = 0 \times M_2$.

We conclude that $M' = M$ and $M'' = 0$, and all this shows that the image of p^* is zero and hence $\omega = 0$. This is false.

We have deduced a contradiction, therefore $2\pi i$ cannot be algebraic. $\qquad\square$

10.2 Transcendence of Logarithms

We now turn to logarithms of algebraic numbers. For $\alpha \in \overline{\mathbb{Q}}^*$, we have

$$\int_1^\alpha \frac{dz}{z} = \log \alpha$$

(with the branch depending on the choice of the path from 1 to α). This is obviously an (incomplete) period of an algebraic variety. In order to apply our theorems directly, we identify it with the period of a 1-motive.

We put

$$M(\alpha) = [\mathbb{Z} \xrightarrow{1 \mapsto \alpha} \mathbb{G}_m].$$

This is often called *Kummer motive* in the literature. If α is a root of unity, this is the motive $M = M_1 \times M_2$ from above. Otherwise the extension

$$0 \to [0 \to \mathbb{G}_m] \to M(\alpha) \to [\mathbb{Z} \to 0] \to 0$$

is non-trivial. By definition, $M(\alpha)^\natural$ is an extension of $\mathbb{G}_m$ by $\mathbb{G}_a = \mathrm{Hom}(\mathbb{Z}, \mathbb{G}_a)$. By Theorem 4.3, $M(\alpha)^\natural$ is canonically isomorphic to $\mathbb{G}_a \times \mathbb{G}_m$. Alternatively, we can also get the splitting by applying the functor $(\cdot)^\natural$ to the natural sequence of motives above. We put

$$\omega = (0, dz/z) \in V_{\mathrm{dR}}^\vee(M(\alpha)) = \mathrm{coLie}(\mathbb{G}_a \times \mathbb{G}_m).$$

The singular realisation of $M(\alpha)$ is

$$\exp_{\mathbb{G}_m}^{-1}(\alpha^{\mathbb{Z}}) \subset \mathrm{Lie}(\mathbb{G}_m)^{\mathrm{an}},$$

which gives

$$V_{\text{sing}}(M(\alpha)) = \langle I(\gamma_1), I(\gamma(\alpha)) \rangle_{\mathbb{Q}} \subset \text{Lie}(\mathbb{G}_m)^{\text{an}},$$

with γ_1 as before the positively oriented loop around 0 and $\gamma(\alpha)$ a path from 1 to α in $\mathbb{G}_m^{\text{an}} = \mathbb{C}^*$. Note that the basis depends on the choice of path $\gamma(\alpha)$, but the lattice does not. We introduce $\sigma(\alpha) = I(\gamma(\alpha)) \in V_{\text{sing}}(M(\alpha))$. Then

$$\int_{\sigma(\alpha)} \frac{dz}{z} = \int_{\gamma(\alpha)} \frac{dz}{z} = \log \alpha,$$

again with the choice of logarithm determined by the choice of path. The calculation of the periods for $M(\alpha)$ uses the canonical embedding

$$V_{\text{sing}}(M(\alpha)) \subset \text{Lie}(M(\alpha)^{\natural}) = \text{Lie}(\mathbb{G}_a)^{\text{an}} \times \text{Lie}(\mathbb{G}_m)^{\text{an}} = \mathbb{C} \times \mathbb{C}$$

given by

$$\sigma_1 \to (0, \sigma_1), \quad \sigma(\alpha) \mapsto \sigma(\alpha)^{\natural} = (1, \sigma(\alpha)) = I(\gamma(\alpha)^{\natural}),$$

with $\gamma(\alpha)^{\natural}(s) = (s, \gamma(\alpha)(s)) \in M(\alpha)^{\natural,\text{an}} = \mathbb{C} \times \mathbb{C}^*$. The period $\omega(\sigma(\alpha))$ is defined by applying the cotangent vector ω to the tangent vector $\sigma(\alpha)^{\natural}$. Hence the period pairing gives

$$\omega(\sigma(\alpha)) = \int_{\gamma(\alpha)^{\natural}} \frac{dz}{z} = \int_{\gamma(\alpha)} \frac{dz}{z} = \log \alpha.$$

Corollary 10.2 (Transcendence of logarithms, Lindemann 1882). *For $\alpha \neq 1$ algebraic, $\log \alpha$ is transcendental, independent of the choice of the branch of the logarithm.*

Proof If α is a root of unity, then $\log \alpha$ is a rational multiple of $2\pi i$, whose transcendence we have already established. From now on let α not be a root of unity and as a consequence $M(\alpha)$ is non-split. Let ω and $\sigma(\alpha)$ be as above and assume that $\int_{\sigma(\alpha)} \omega$ is algebraic. By Theorem 9.11 there is a decomposition

$$\omega = \psi + \phi$$

with $\int_{\sigma(\alpha)} \psi = 0$ and such that the image of ϕ vanishes in $V_{\text{dR}}^{\vee}(\mathbb{G}_m)$. We first concentrate on ψ. An application of Theorem 9.7 to ψ gives a short exact sequence

$$0 \to M' \xrightarrow{i} M(\alpha) \xrightarrow{p} M'' \to 0$$

such that $\sigma(\alpha)$ is in the image of i_* and ψ is in the image of p^*. There are only three possibilities for M':

$$0, \quad [0 \to \mathbb{G}_m], \quad M(\alpha).$$

We exclude $M' = 0$ because $\sigma(\alpha) \neq 0$ and $M' = [0 \to \mathbb{G}_m]$ because $\sigma(\alpha) \notin V_{\text{sing}}([0 \to \mathbb{G}_m])$. This shows that $M' = M(\alpha)$ and implies that $M'' = 0$ and we conclude that $\psi = 0$. As a consequence $\omega = \phi$ vanishes when mapped to $V_{\text{dR}}^{\vee}([0 \to \mathbb{G}_m])$. But actually this image is dz/z, so we have a contradiction.

□

10.3 Hilbert's Seventh Problem

In his seventh problem Hilbert asked whether

$$\alpha, \beta, \alpha^{\beta}$$

can all be algebraic unless $\alpha = 0, 1$ or β rational. In other words:

(GS) $\qquad\qquad\qquad \alpha, \beta, \alpha^{\beta} \in \overline{\mathbb{Q}} \Rightarrow \alpha = 0, 1$ or $\beta \in \mathbb{Q}$.

There is a logarithmic version of (GS): let $\alpha, \gamma \in \overline{\mathbb{Q}}^{*}$, $\log \alpha$ and $\log \gamma$ choices of branches of logarithm.

(B) $\qquad\qquad$ $\log \alpha$ and $\log \gamma$ are $\overline{\mathbb{Q}}$-linearly dependent

$\qquad\qquad\qquad \Rightarrow \log \alpha$ and $\log \gamma$ are $\mathbb{Q}$-linearly dependent.

We think of (GS) as the Gelfond–Schneider version of the implication and of (B) as the Baker version.

Note that the converse implication of (B) is obvious. However, the converse implication of (GS) fails in the case $\alpha = 1$ if we do not use the principal branch of logarithm in the definition of α^{β}. The problem disappears when restricting to real numbers.

Lemma 10.3. *The implications (GS) for all α, β and γ and (B) for all α and γ are equivalent.*

Proof We assume (GS). Let α, γ be non-zero algebraic numbers such that $\log \alpha$ and $\log \gamma$ are $\overline{\mathbb{Q}}$-linearly dependent. By assumption there are $u, v \in \overline{\mathbb{Q}}^{*}$ such that

$$u \log \alpha + v \log \gamma = 0.$$

Without loss of generality, we may assume that $\alpha \neq 1$. Indeed, if $\alpha = 1$ and $\log \alpha = 2\pi n$ for $n \in \mathbb{Z}$, we consider instead $\alpha = \gamma$ and different branches of $\log \gamma$.

We introduce $\beta = u/v$. As a consequence of the linear relation from above, we find that $\alpha^{\beta} = \exp(\beta \log \alpha) = \gamma^{-1} \in \overline{\mathbb{Q}}^{*}$. By (GS), this implies $\beta \in \mathbb{Q}$.

Conversely, we assume that (B). We take $\alpha, \beta, \gamma \in \overline{\mathbb{Q}}$, $\alpha \neq 0, 1$ such that $\gamma = \alpha^\beta$ and then

$$\log \gamma = \beta \log \alpha + 2\pi i n$$

for some choice of logarithms and an appropriate $n \in \mathbb{Z}$. We replace the choice of branch of the logarithm for γ such that $n = 0$. Then $\log \gamma$ and $\log \alpha$ are $\overline{\mathbb{Q}}$-linearly dependent. By (B) this implies $\beta \in \mathbb{Q}$. $\qquad\square$

Theorem 10.4 (Gelfond–Schneider 1934). *Implication (B) holds.*

Proof We switch notation from (B) and write β instead of γ. Having fixed $\alpha, \beta \in \overline{\mathbb{Q}}^*$ and branches of logarithm $\log \alpha$, $\log \beta$ such that the numbers are $\overline{\mathbb{Q}}$-linearly dependent, we find $a, b \in \overline{\mathbb{Q}}^*$ such that

$$a \log \alpha + b \log \beta = 0.$$

In order to show that a and b are $\mathbb{Q}$-linearly dependent, we consider the 1-motive $M = [\mathbb{Z} \to \mathbb{G}_m^2]$ with structure morphism $1 \mapsto (\alpha, \beta)$. The motive is split, i.e. isogenous to $[\mathbb{Z} \to 0] \oplus [0 \to \mathbb{G}_m^2]$ if both α and β are roots of unity. In this case $\log \alpha$ and $\log \beta$ are rational multiples of $2\pi i$, hence linearly dependent. This case is excluded from now on.

Similar to the case of transcendence of logarithms, we have $M^\natural = \mathbb{G}_a^1 \times \mathbb{G}_m^2$ and hence $V_{\mathrm{dR}}^\vee(M) = \mathrm{coLie}(M^\natural)$ has the basis $(dt, 0, 0), (0, dz_1/z_1, 0), (0, 0, dz_2/z_2)$. We put $\omega = (0, a\, dz_1/z_1, b\, dz_2/z_2)$.

Our next step is to compute $T_{\mathrm{sing}}(M)$. As M is non-split, $T_{\mathrm{sing}}(M)$ is a subset of $\mathrm{Lie}(\mathbb{G}_m^2)^{\mathrm{an}}$. Recall that $\log \alpha \cdot d/dz_1 = I(\gamma(\alpha))$, where $\gamma(\alpha)$ is a suitable path in $\mathbb{G}_m^{\mathrm{an}}$ from 1 to α. The same relation holds for β. Let $\gamma : [0, 1] \to \mathbb{G}_m^{\mathrm{an}} \times \mathbb{G}_m^{\mathrm{an}}$ be given by $\gamma(t) = (\gamma(\alpha)(t), \gamma(\beta)(t))$ and put $\sigma = I(\gamma) \in T_{\mathrm{sing}}(M) \subset \mathrm{Lie}(\mathbb{G}_m^2)^{\mathrm{an}}$. By construction,

$$\omega(\sigma) = \int_{\gamma(\alpha)} a \frac{dz_1}{z_1} + \int_{\gamma(\beta)} b \frac{dz_2}{z_2} = a \log \alpha + b \log \beta = 0,$$

and Theorem 9.7 furnishes a short exact sequence

$$0 \to M_1 \to M \to M_2 \to 0$$

such that σ is induced from M_1 and ω from $M_2 = [\mathbb{Z}^s \to \mathbb{G}_m^t]$ with $s \leq 1$ and $t \leq 2$.

If $t = 2$, then the surjection $M \to M_2$ is the identity on the torus part. The push-forward of σ is given by $(\log \alpha \cdot d/dz_1, \log \beta \cdot d/dz_2)$. Hence both vanish. This implies that $\alpha = \beta = 1$, a case we had excluded.

The case $t = 0$ does not occur because ω is not a pull-back from a motive of the form $[\mathbb{Z}^s \to 0]$.

We are left with the case $t = 1$. The torus part of the map $M \to M_2$ is given by $(x, y) \to x^n y^m$ for $n, m \in \mathbb{Z}$. The induced map on Lie algebras maps $(d/dt, \log \alpha \cdot d/dz_1, \log \beta \cdot d/dz_2)$ to $(n \log \alpha + m \log \beta) d/dz$. This image vanishes and gives the linear dependence we were looking for. □

What we have just seen is a motivic reformulation of Gelfond's proof based on $\mathbb{G}_m \times \mathbb{G}_m$. In contrast, Schneider's argument uses $\mathbb{G}_a \times \mathbb{G}_m$ but does not have a translation to our language. One would need a modification of the Analytic Subgroup Theorem, which would be desirable.

The same arguments also apply to more than two numbers, which leads to the following result.

Theorem 10.5 (Baker 1967). *Take $\alpha_1, \ldots, \alpha_n \in \overline{\mathbb{Q}}^*$. If $\log \alpha_1, \ldots, \log \alpha_n$ are $\overline{\mathbb{Q}}$-linearly dependent, then they are $\mathbb{Q}$-linearly dependent.*

We even have

$$\mathrm{rk}\langle \alpha_1, \ldots, \alpha_n \rangle_{\mathbb{Z}} = \dim_{\overline{\mathbb{Q}}} \langle \log \alpha_1, \ldots, \log \alpha_n, 2\pi i \rangle_{\overline{\mathbb{Q}}} / 2\pi i \overline{\mathbb{Q}}$$

for any choice of branches of logarithms.

Remark 10.6. In the literature we often find formulations with $\alpha_1, \ldots, \alpha_n$ multiplicatively independent. The above is the correct version that also allows roots of unity or even repetitions with different choices of branch of logarithm. We will discuss in more detail later (see Chapter 16) that the space of periods of the third kind with respect to non-closed paths is only well defined up to other types of periods.

10.4 Abelian Periods for Closed Paths

Another important case involves periods of abelian varieties in the classical sense.

Corollary 10.7 (Wüstholz [Wüs87]). *Let A be an abelian variety, $\omega \in \Omega^1(A)$ and γ a closed path on A^{an}. Then*

$$\int_{\gamma} \omega$$

is either 0 or transcendental.

Proof Consider $M = [0 \to A]$. Its de Rham realisation is $\mathrm{coLie}(A^{\natural})^{A^{\natural}} \supset \mathrm{coLie}(A)^A$. All global differential forms on A are A-invariant, hence ω defines an element of $V_{\mathrm{dR}}^{\vee}(M)$. Its singular realisation is by definition the kernel of $\exp_A \colon \mathrm{Lie}(A)^{\mathrm{an}} \to A^{\mathrm{an}}$. Let $\sigma \in V_{\mathrm{sing}}(A)$ be the element such that the image of a path from 0 to σ under $\exp_A$ is equal to γ. Then

$$\int_\sigma \omega = \int_\gamma \omega.$$

Assume that the period is algebraic. An application of Theorem 9.11 gives $\omega = \phi + \psi$, with $\int_\sigma \psi = 0$ and ϕ in the kernel of the restriction to the group part of M. But M is equal to its group part, implying that $\phi = 0$.   □

11

On Non-closed Elliptic Periods

The computation of the dimension of the period space is a classical problem and has been studied in various cases by many authors. In this chapter, we concentrate on the case of a non-classical elliptic 1-motive. For instance, we deduce the first examples of periods which were not known to be transcendental. At this point everything will be formulated in terms of 1-motives. For the translation to periods of the first, second and third kind on curves, see Chapters 14 and 18.

The dimension formula is a special case of the generalised Baker theory in Part IV. We give a direct proof that should be understood as a warm-up for the considerably more complicated general case. This special result will not be needed later on.

11.1 The Setting

Let $A = E$ be an elliptic curve, $0 \to \mathbb{G}_m \to G \to E \to 0$ a non-trivial extension (which remains non-split up to isogeny) and $P \in G(\overline{\mathbb{Q}})$ a point whose image in $E(\overline{\mathbb{Q}})$ is not a torsion. We consider the 1-motive

$$M = [\mathbb{Z} \to G],$$

with 1 mapping to P. We denote by $\delta(M)$ the dimension of the $\overline{\mathbb{Q}}$-vector space $\mathcal{P}\langle M \rangle$ generated by the periods of M in $\mathbb{C}$.

We start by choosing bases in the singular and de Rham cohomology respecting the weight filtration. The inclusions

$$[0 \to \mathbb{G}_m] \hookrightarrow [0 \to G] \hookrightarrow M$$

with cokernels $[0 \to E]$ and $[\mathbb{Z} \to 0]$, respectively, lead to a filtration

$$V_{\text{sing}}(\mathbb{G}_m) \subset V_{\text{sing}}(G) \subset V_{\text{sing}}(M).$$

Extend a basis σ for $V_{\text{sing}}(\mathbb{G}_m)$ by γ_1, γ_2 to a basis of $V_{\text{sing}}(G)$ and further by λ to a basis of the whole space. Then their images $\bar{\gamma}_1, \bar{\gamma}_2$ in E form a basis of $V_{\text{sing}}(E)$ and $\bar{\lambda}$ forms a basis of $V_{\text{sing}}([\mathbb{Z} \to 0])$.

For the de Rham realisation consider the cofiltration

$$M \twoheadrightarrow [\mathbb{Z} \to E] \twoheadrightarrow [\mathbb{Z} \to 0]$$

with kernels $[0 \to \mathbb{G}_m]$ and $[0 \to E]$, respectively. They lead by pull-back of forms to a filtration

$$V_{\text{dR}}^{\vee}(M) \supset V_{\text{dR}}^{\vee}([\mathbb{Z} \to E]) \supset V_{\text{dR}}^{\vee}([\mathbb{Z} \to 0]).$$

Extend a basis u of $V_{\text{dR}}^{\vee}([\mathbb{Z} \to 0])$ by ω, η to a basis of $V_{\text{dR}}^{\vee}([\mathbb{Z} \to E])$ and by ξ to a basis of $V_{\text{dR}}^{\vee}(M)$. The images $\bar{\omega}, \bar{\eta}$ of ω, and η are a basis of $V_{\text{dR}}^{\vee}(E)$, and the image $\bar{\xi}$ of ξ is a basis of $V_{\text{dR}}^{\vee}(\mathbb{G}_m)$. The period space $\mathcal{P}\langle M \rangle$ is spanned by the numbers obtained by pairing our basis vectors.

The pairing of an element of $V_{\text{sing}}(M)$ coming from a subobject with an element of $V_{\text{dR}}^{\vee}(M)$ coming from the corresponding quotient is zero, as we know. Applying this observation to the two filtrations from above gives $u(\sigma) = \omega(\sigma) = u(\gamma_1) = u(\gamma_2) = \omega(\sigma) = 0$. Further, one sees that $u(\lambda) = 1$ and $\xi(\sigma) = 2\pi i$ (at least after scaling). Taking this together gives a period matrix of the shape

$$\begin{pmatrix} 2\pi i & \xi(\gamma_1) & \xi(\gamma_2) & \xi(\lambda) \\ 0 & \omega(\gamma_1) & \omega(\gamma_2) & \omega(\lambda) \\ 0 & \eta(\gamma_1) & \eta(\gamma_2) & \eta(\lambda) \\ 0 & 0 & 0 & 1 \end{pmatrix}.$$

The calculation of the dimension of the associated space of periods needs to distinguish between two cases, the CM-case and the non-CM-case. We deal with each of the two cases separately.

11.2 Without CM

In the case when there is no complex multiplication, the non-CM-case, the endomorphism algebra $\text{End}(E) = \mathbb{Z}$ is trivial.

Proposition 11.1. *Let M be as just described. Then*

$$\delta(M) = 11.$$

This will also be a corollary of the general theory in Part IV. The deduction of the corollary is explained in Example 15.4 and its continuation in Example 17.16.

Direct proof. We use the notation fixed above. It has to be shown that all entries of the period matrix are $\overline{\mathbb{Q}}$-linearly independent. If not, there is a relation

$$a\,2\pi i + \sum_{i=1}^{2}(b_i\,\xi(\gamma_i) + c_i\,\omega(\gamma_i) + d_i\,\eta(\gamma_i)) + e\,\xi(\lambda) + f\,\omega(\lambda) + g\,\eta(\lambda) + h = 0$$

with $a, b_1, b_2, c_1, c_2, d_1, d_2, e, f, g, h \in \overline{\mathbb{Q}}$. We consider the motive

$$\widetilde{M} = [0 \to \mathbb{G}_m] \times [0 \to G]^2 \times M = [0^3 \times \mathbb{Z} \to \mathbb{G}_m \times G^3]$$

together with

$$\widetilde{\gamma} = (\sigma, \gamma_1, \gamma_2, \lambda) \in V_{\mathrm{sing}}(\widetilde{M}),$$
$$\widetilde{\omega} = (a\xi, b_1\xi + c_1\omega + d_1\eta, b_2\xi + c_2\omega + d_2\eta, e\xi + f\omega + g\eta + hu) \in V_{\mathrm{dR}}^{\vee}(\widetilde{M}).$$

Then $\widetilde{\omega}(\widetilde{\gamma}) = 0$. By the Analytic Subgroup Theorem for 1-motives, see Theorem 9.7, there is a short exact sequence

$$0 \to M_1 \xrightarrow{i} \widetilde{M} \xrightarrow{p} M_2 \to 0$$

of 1-motives $M_1 = [L_1 \to G_1]$ and $M_2 = [L_2 \to G_2]$ with $\widetilde{\gamma} = i_*\gamma_1$ for some $\widetilde{\gamma}_1 \in V_{\mathrm{sing}}(M_1)$ and $\widetilde{\omega} = p^*\omega_2$ for some $\omega_2 \in V_{\mathrm{dR}}^{\vee}(M_2)$.

Let A_2 be the abelian part of M_2. We want to show that $A_2 = 0$. Assuming $A_2 \neq 0$, we choose a non-zero surjective map $\widetilde{\kappa}\colon \widetilde{M} \to [L \to E]$ which factors as $\kappa_2 \circ p$ through p. As L is a quotient of $\mathbb{Z}$, there are (up to isogeny) only two possibilities, namely $L = 0$ or $L = \mathbb{Z}$. The map $\widetilde{\kappa}$ factors via

$$\kappa\colon [0^3 \times \mathbb{Z} \to 0 \times E^3] \to [L \to E].$$

On the abelian part it is given by a vector $(0, n, m, k)$ with $n, m, k \in \mathrm{End}(E) = \mathbb{Z}$. Note that there is no complex multiplication. We deduce that $L = k\mathbb{Z}$, which is non-zero if and only if $k \neq 0$. Since $\widetilde{\kappa}_*\widetilde{\gamma} = \kappa_{2*} \circ p_* \circ i_*\widetilde{\gamma}_1 = 0$, we deduce that

$$0 = \kappa_*\widetilde{\gamma} = n\gamma_1 + m\gamma_2 + k\lambda \in V_{\mathrm{sing}}([L_\kappa \to E]).$$

The elements γ_1, γ_2 (and if $k \neq 0$ also λ) are linearly independent in the vector space $V_{\mathrm{sing}}([L \to E])$, which implies that $n = m = k = 0$. This contradicts the non-triviality of $\widetilde{\kappa}$ and proves that $A_2 = 0$.

In conclusion, we have $M_2 = [L_2 \to \mathbb{G}_m^r]$ for some $0 \leq r \leq 4$. The group part of the morphism of motives $p\colon \widetilde{M} \to M_2$ has the form

$$\theta\colon \mathbb{G}_m \times G^3 \twoheadrightarrow \mathbb{G}_m^r.$$

Its components $G \to \mathbb{G}_m^r$ have to vanish since G is non-split. The surjectivity of θ implies that $r \leq 1$ with $\theta = (?, 0, 0, 0)$ either 0 or the projection to the factor $\mathbb{G}_m$.

This gives us a lot of information on $\widetilde{\omega}$. Recall that $\widetilde{\omega} = p^*\omega_2$ for some $\omega_2 \in V_{\mathrm{dR}}^\vee(M_2)$. We have the commutative diagram

$$
\begin{array}{ccc}
[0 \to \mathbb{G}_m \times G^3] & \xrightarrow{\ \theta\ } & [0 \to \mathbb{G}_m^r] \\
\downarrow & & \downarrow \\
\widetilde{M} & \xrightarrow{\ \ p\ \ } & M_2.
\end{array}
$$

Hence the pull-back of $\widetilde{\omega}$ to $V_{\mathrm{dR}}^\vee\left(\mathbb{G}_m \times G^3\right)$ is concentrated in the first component, which gives

$$b_1\xi + c_1\omega + d_1\eta = 0,$$
$$b_2\xi + c_2\omega + d_2\eta = 0,$$
$$e\xi + f\omega + g\eta = 0.$$

As the three classes ξ, ω, η are linearly independent, we get vanishing coefficients

$$b_1 = b_2 = c_1 = c_2 = d_1 = d_2 = e = f = g = 0.$$

We are left with the case $\widetilde{\omega} = (a\xi, 0, 0, hu)$. If $a \neq 0$, then from the period relation also $h \neq 0$, and conversely. Assume this is the case. As $\widetilde{\omega} = p^*\omega_2$, it follows that the group part of $\widetilde{M} \to M_2$ is the projection to the first factor and $M_2 = \mathbb{G}_m \times [\mathbb{Z} \to 0]$ in this situation. The kernel M_1 is equal to $0 \times G^3$. However, $\widetilde{\gamma}$ is induced from M_1 and therefore of the form $(0, \dots)$. This contradicts $\sigma \neq 0$. Hence we must have $a = 0$ and $h = 0$. $\qquad\qquad\square$

11.3 The CM-Case

We turn to the CM-case when $\mathrm{End}(E)_\mathbb{Q} = \mathbb{Q}(\tau)$ is an imaginary quadratic extension of $\mathbb{Q}$.

The action of $\mathbb{Q}(\tau)$ induces new relations between the entries of the period matrix. This is well known in the language of periods of curves. We take the point of view of 1-motives instead.

The singular realisation $V_{\mathrm{sing}}(E)$ is of dimension 2 as a $\mathbb{Q}$-vector space because E has genus 1. As a $\mathbb{Q}(\tau)$-vector space, it has dimension 1. For any non-zero γ in $V_{\mathrm{sing}}(E)$, the pair $(\gamma, \tau_*\gamma)$ is a $\mathbb{Q}$-basis of $V_{\mathrm{sing}}(E)$. After extension of scalars to $\mathbb{C}$, the operation is still semi-simple, meaning that $V_{\mathrm{sing}}(E)_\mathbb{C}$ decomposes as a sum of two τ_*-eigenspaces with complex conjugate eigenvalues. The dual operation τ^* induced on $V_{\mathrm{sing}}(E)^\vee$ has the same set of eigenvalues.

This can also be described on the de Rham realisation. We calculate $V_{\mathrm{dR}}^{\vee}(E) = \mathrm{coLie}(E^{\natural}) = \Omega^{1}(E^{\natural})^{E^{\natural}}$ by looking at

$$0 \to H^{1}(E, \mathcal{O})^{\vee} \to E^{\natural} \to E \to 0.$$

By Theorem 8.16, $V^{\vee}(E)$ carries a Hodge structure. (This is the 1-motivic incarnation of the Hodge decomposition of $H_{\mathrm{dR}}^{1}(E)$.) It is explicitly given by

$$F^{1}V_{\mathrm{dR}}^{\vee}(E) = \Omega^{1}(E).$$

This $\overline{\mathbb{Q}}$-subspace is invariant under τ^{*}, hence it is one of the eigenspaces. After extension of scalars to $\mathbb{C}$, we even get the following decomposition by Hodge theory:

$$V_{\mathrm{dR}}^{\vee}(E)_{\mathbb{C}} = F^{1}V_{\mathrm{dR}}^{\vee}(E)_{\mathbb{C}} \oplus \overline{F^{1}V_{\mathrm{dR}}^{\vee}(E)_{\mathbb{C}}}.$$

Everything is stable under the τ^{*}-operation, so we have identified the eigenspace description. By definition, the complex number τ corresponds to the unique endomorphism of E which operates as multiplication by τ on the Lie algebra and on the dual $\Omega^{1}(E)$, so the latter is the τ-eigenspace. The eigenvalues are simply τ and $\bar{\tau}$. Since τ^{*} acts on the $\overline{\mathbb{Q}}$-vector space $V_{\mathrm{dR}}^{\vee}(E)$, its eigenvectors are in $V_{\mathrm{dR}}^{\vee}(E)$. Let ω', ω'' be a basis of τ^{*}-eigenvectors of $V_{\mathrm{dR}}^{\vee}(E)$, which has the property that

$$\tau^{*}\omega' = \tau \cdot \omega', \quad \tau^{*}\omega'' = \bar{\tau} \cdot \omega''.$$

Corollary 11.2. *In the basis* $(\gamma, \tau_{*}\gamma)$ *of* $V_{\mathrm{sing}}(E)$ *and* ω', ω'' *of* $V_{\mathrm{dR}}^{\vee}(E)$, *the period relations for E can be expressed as*

$$\omega'(\tau_{*}\gamma) = (\tau^{*}\omega')(\gamma) = \tau\omega'(\gamma),$$
$$\omega''(\tau_{*}\gamma) = (\tau^{*}\eta)(\gamma) = \bar{\tau}\omega''(\gamma).$$

Proposition 11.3. *Let M be the motive introduced in Section 11.1 with* $\mathrm{End}_{\mathbb{Q}}(E) = \mathbb{Q}(\tau)$. *Then*

$$\delta(M) = 9.$$

Direct proof.

$$\widetilde{M} = [0 \to \mathbb{G}_{m}] \times [0 \to G] \times M.$$

We go through the proof in the non-CM-case and make the necessary changes. Let $\sigma, \gamma_{1}, \gamma_{2}, \lambda$ be as in Section 11.1. We choose a little more carefully $\gamma_{2} = \tau_{*}\gamma_{1}$ and then

$$\omega(\tau_{*}\gamma_{1}) = (\tau^{*}\omega)(\gamma_{1}) = a\,\omega(\gamma_{1}) + b\,\eta(\gamma_{1}) + c\,u(\gamma_{1}),$$
$$\eta(\tau_{*}\gamma) = (\tau^{*}\eta)(\gamma_{1}) = d\,\omega(\gamma_{1}) + e\,\eta(\gamma_{1}) + f\,u(\gamma_{1}),$$

with $a, b, c, d, e, f \in \overline{\mathbb{Q}}$. Indeed, in the choice of basis used in the corollary leads to $a = \tau$, $b = d = 0$, $e = \bar{\tau}$; but we do not need the special shape. Note that the argument does not apply to $\xi(\gamma_2)$ because $\tau^*\xi$ is not defined, or, in other words, would relate not to M but to a different 1-motive.

It remains to show that $2\pi i, \xi(\gamma_1), \omega(\gamma_1), \eta(\gamma_1), \xi(\gamma_2), \xi(\lambda), \omega(\lambda), \eta(\lambda), 1$ are linearly independent. If they are not, there is a linear relation as in the first case, but omitting the summands for $\omega(\gamma_2)$ and $\eta(\gamma_2)$, so $c_2 = d_2 = 0$. We consider the motive

$$\widetilde{M} = [0 \to \mathbb{G}_m] \times [0 \to G] \times M$$

and $\widetilde{\gamma}, \widetilde{\omega}$ analogously to before. Again this gives M_1, M_2. Assume that $A_2 \neq 0$ and choose $\kappa_2, L_\kappa, \kappa$ as in the first case. The composition

$$\widetilde{A} = 0 \times E^2 \to E$$

is now given by a vector $(0, n, k)$ of elements of $\mathrm{End}(E) \subset \mathbb{Q}(\tau)$. The rest of the argument is the same as in the non-CM-case. $\qquad\square$

11.4 Transcendence

As a simple corollary of the explicit dimension computation, we also deduce the transcendence of periods of our M. We concentrate on the case where transcendence is not a simple consequence of the Analytic Subgroup Theorem. In the language of Chapter 16, this refers to a period of the third kind with respect to a non-closed path.

Corollary 11.4. *Let* $M = [\mathbb{Z} \to G]$, *and let* $\sigma, \gamma_1, \lambda \in V_{\mathrm{sing}}(M)$, $\omega, \eta, \xi \in V_{\mathrm{dR}}^\vee(M)$ *be as in Section 11.1. Then the periods*

$$\omega(\gamma_1), \omega(\lambda), \eta(\gamma_1), \eta(\lambda), \xi(\sigma) = 2\pi i, \xi(\gamma_1), \xi(\lambda)$$

are transcendental.

Proof These elements agree with those from the proofs of Proposition 11.1 (non-CM-case) and Proposition 11.3. In both cases, 1 appears as a period and we explicitly proved linear independence of the two periods. $\qquad\square$

Remark 11.5. The same transcendence result already appears in [Wüs21]. For $\omega(\gamma_1)$, $\eta(\gamma_1)$ this means the transcendence of periods and quasi-periods of elliptic curves. These are old results of Siegel [Sie32] and Schneider [Sch34b, Sch34a]. The transcendence of $\omega(\lambda)$, $\eta(\lambda)$ and $\xi(\gamma_1)$ can be deduced from the Analytic Subgroup Theorem without the detour through 1-motives.

We may also deduce the same result more directly from the Analytic Subgroup Theorem for motives. We show the most interesting case $\xi(\lambda)$ as an example. The other cases can be treated in the same way.

Second proof. Assume that $\xi(\lambda)$ is algebraic. We apply the transcendence criterion given in Theorem 9.11 and write accordingly $\xi = \phi + \psi$ such that $\psi(\lambda) = 0$ and the image of ϕ in $V_{\mathrm{dR}}^{\vee}(G)$ vanishes. A fortiori the image of ϕ in $V_{\mathrm{dR}}^{\vee}(\mathbb{G}_m)$ vanishes. This implies that ψ is simply an alternative choice for ξ. It suffices to consider the case

$$\xi(\lambda) = 0.$$

By the Analytic Subgroup Theorem for 1-motives, Theorem 9.7, there is a short exact sequence

$$0 \to M' \xrightarrow{i} M \xrightarrow{p} M'' \to 0$$

such that $\lambda = i_* \lambda'$, $\xi = p^* \xi''$.

The simple constituents of our M are $\mathbb{G}_m, E, [\mathbb{Z} \to 0]$. As the image of λ in $V_{\mathrm{sing}}([\mathbb{Z} \to 0])$ and the image of ξ in $V_{\mathrm{dR}}^{\vee}(\mathbb{G}_m)$ are bases, we know that $[\mathbb{Z} \to 0]$ must be a constituent of M' and $\mathbb{G}_m$ a constituent of M''. Hence there are only two possible shapes for M':

$$[\mathbb{Z} \to 0], \quad [\mathbb{Z} \to E].$$

In the first case, $M' = [\mathbb{Z} \to 0]$ and the inclusion i is a section of the natural surjection $M \to [\mathbb{Z} \to 0]$ and even of $[\mathbb{Z} \to E] \to [\mathbb{Z} \to E]$. This contradicts the choice of P in the definition of M.

In the second case, $M'' = \mathbb{G}_m$. The projection p is a section of the natural inclusion $\mathbb{G}_m \to M$ and even of $\mathbb{G}_m \to G$. This contradicts the choice of G in the definition of M.

The two contradictions lead to $\xi(\lambda) \neq 0$. $\qquad\qquad\square$

PART THREE

PERIODS OF ALGEBRAIC VARIETIES

12

Periods of Algebraic Varieties

We give an alternative description of the set of periods of 1-motives as periods of the first cohomology of algebraic varieties defined over the algebraic closure $\overline{\mathbb{Q}}$ of $\mathbb{Q}$. This needs to fix an embedding $\overline{\mathbb{Q}} \to \mathbb{C}$.

12.1 Spaces of Cohomological 1-Periods

The basic objects are triples (X, Y, i), where X is a $\overline{\mathbb{Q}}$-variety, Y a closed subvariety and $i \in \mathbb{N}_0$. Let $H^i_{\mathrm{dR}}(X, Y)$ be the relative de Rham cohomology of such a triple (see Section 3.2) and $H^i_{\mathrm{sing}}(X, Y; \mathbb{Q})$ the relative singular cohomology (see Section 3.1). The first is a $\overline{\mathbb{Q}}$-vector space, the second a $\mathbb{Q}$-vector space. After base change to the complex numbers they become naturally isomorphic via the period isomorphism ϕ (see Section 3.3). In good cases, the period isomorphism can be explicitly described as integration of closed differential forms over singular cycles.

In Chapter 7 we introduced the category $(\overline{\mathbb{Q}}, \mathbb{Q})$-Vect with objects of the form $V = (V_{\overline{\mathbb{Q}}}, V_{\mathbb{Q}}, \phi)$, where $V_{\overline{\mathbb{Q}}}$ is a finite-dimensional $\overline{\mathbb{Q}}$-vector space, $V_{\mathbb{Q}}$ a finite-dimensional $\mathbb{Q}$-vector space and $\phi \colon V_{\mathbb{Q}} \otimes_{\mathbb{Q}} \mathbb{C} \to V_{\overline{\mathbb{Q}}} \otimes_{\overline{\mathbb{Q}}} \mathbb{C}$ an isomorphism.

Definition 12.1. For algebraic varieties $X \supset Y$ over $\overline{\mathbb{Q}}$ and $i \in \mathbb{N}_0$, we denote by $H^i(X, Y) \in (\overline{\mathbb{Q}}, \mathbb{Q})$-Vect the triple $(H^i_{\mathrm{dR}}(X, Y), H^i_{\mathrm{sing}}(X, Y; \mathbb{Q}), \phi)$.

The assignment $(X, Y, i) \to H^i(X, Y)$ is natural for morphisms of pairs $(X, Y) \to (X', Y')$. For every triple $X \supset Y \supset Z$ there are connecting morphisms

$$\partial \colon H^i(Y, Z) \to H^{i+1}(X, Y)$$

which are morphisms in $(\overline{\mathbb{Q}}, \mathbb{Q})$-Vect.

111

Definition 12.2. For $H^i(X, Y)$ as defined above, the *set of period numbers* $\mathcal{P}(X, Y, i)$ is the image of the period pairing

$$H^i_{dR}(X, Y) \times H^{\text{sing}}_i(X, Y; \mathbb{Q}) \to \mathbb{C}.$$

The set of i-periods $\mathcal{P}^i$ is the union of the $\mathcal{P}(X, Y, i)$ for all X and Y.

In this book we are primarily interested in the case $i = 1$.

Example 12.3. We have $\mathcal{P}^0 = \overline{\mathbb{Q}}$ because $H^0(X, Y)$ depends only on the connected components of X and Y.

Lemma 12.4. *For all i, the set $\mathcal{P}^i$ is a $\overline{\mathbb{Q}}$-subspace of $\mathbb{C}$. We have $\mathcal{P}^i \subset \mathcal{P}^{i+1}$.*

Proof A sum of two periods of $H^i(X, Y)$ and $H^i(X', Y')$ can be realised as a period of $H^i(X \sqcup X', Y \sqcup Y')$. The set is stable under multiplication by numbers in $\overline{\mathbb{Q}}$ because $H^i_{dR}(X, Y)$ is a $\overline{\mathbb{Q}}$-vector space. Integration of the differential form dt over the path $s \mapsto e^{2\pi i s}$ on $[0, 1]$ gives 1 as a period of $H^1(\mathbb{A}^1, \{0, 1\})$, whence the periods of $H^i(X, Y)$ are contained in (actually equal to) the set of periods of $H^{i+1}(X \times \mathbb{A}^1, Y \times \mathbb{A}^1 \cup X \times \{0, 1\}) \cong H^i(X, Y) \otimes H^1(\mathbb{A}^1, \{0, 1\})$. $\square$

12.2 Periods of Curve Type

Nori showed that every affine algebraic variety admits a filtration by subvarieties defined over $\overline{\mathbb{Q}}$ such that their relative homology is concentrated in a single degree. This 'good filtration' should be seen as an analogue of the skeletal filtration of a simplicial complex or a CW-complex. Indeed, affine algebraic varieties have the homotopy type of a simplicial complex. The surprising insight is the existence of such a filtration by algebraic subvarieties, even over the ground field. This filtration goes into the construction of the category of Nori motives, but it also has immediate consequences for periods. For the general result, see [HMS17, Section 11]. We repeat the argument in our case.

Proposition 12.5. *In the definition of $\mathcal{P}^1$ it suffices to consider $H^1(C, D)$, where C is a smooth affine curve and D a finite collection of points on C.*

Proof Consider the periods of $H^1(X, Y)$ for arbitrary X and Y. We first show that it suffices to deal with affine varieties X. By Jouanolou's trick there is an $\mathbb{A}^n$-torsor $\widetilde{X} \to X$ with $\widetilde{X}$ affine. Let $\widetilde{Y}$ be the preimage of Y in $\widetilde{X}$. Since the map $\widetilde{X} \to X$ is a homotopy equivalence, we have

$$H^1(X, Y) \cong H^1(\widetilde{X}, \widetilde{Y}).$$

Therefore we may without loss of generality assume that X (and consequently also Y) is affine.

Nori's Basic Lemma in [HMS17, Proposition 9.2.3, Corollary 9.2.5] provides our affine varieties with *very good filtrations* by closed subvarieties

$$X_0 \subset X_1 \subset \cdots \subset X_n = X, \quad Y_0 \subset Y_1 \subset \cdots \subset Y_n = Y$$

with $Y_i \subset X_i$. By definition this means that

1. $X_i \smallsetminus X_{i-1}$ is smooth,
2. either $\dim X_i = i$ and $\dim X_{i-1} = i-1$ or $X_i = X_{i-1}$ and the dimension of $\dim X_i$ is less than i,
3. $H^j(X_i, X_{i-1}) = 0$ for $j \neq i$,

and the same holds for Y_i. The boundary maps in the long exact sequence for the triple (X_{i+1}, X_i, X_{i-1}) introduced in Section 3.1 define a complex

$$C(X_*) = [H^0(X_0) \to H^1(X_1, X_0) \to H^2(X_2, X_1) \to \cdots].$$

By [Hat02, Theorem 2.35] its cohomology in degree j agrees with $H^j(X)$. We introduce

$$C(X_*, Y_*) = \operatorname{cone}(C(X_*) \to C(Y_*))[-1],$$

explicitly given by

$$[H^0(X_0) \to H^1(X_1, X_0) \oplus H^0(Y_0) \to H^1(X_2, X_1) \oplus H^1(Y_1, Y_0) \to \cdots].$$

Its cohomology in degree j agrees naturally with $H^j(X, Y)$. As a result, $H^1(X, Y)$ can be identified with a subquotient of

$$H^1(X_1, X_0) \oplus H^0(Y_0).$$

This implies that the periods of $H^1(X, Y)$ are contained in the space of periods of $H^1(X_1, X_0) \oplus H^0(Y_0)$. As discussed in the proof of Lemma 12.4, the periods of $H^0(Y_0)$ can also be seen as periods of $H^1(Y_0 \times \mathbb{A}^1)$, so they are periods of smooth affine curves.

In conclusion, it remains to consider the case where X is an affine curve, Y a finite set of points and, in addition, $X \smallsetminus Y$ is smooth. By normalisation, we resolve the singularities of X. We denote by $\widetilde{X}$ the normalisation and by $\widetilde{Y}$ the preimage of Y in $\widetilde{X}$. By excision, we have

$$H^1(X, Y) \cong H^1(\widetilde{X}, \widetilde{Y}).$$

The curve $\widetilde{X}$ is smooth and affine. □

Definition 12.6. We say that a period is *of curve type* if it is the period of some $H^1(C, D)$, where C is a smooth affine curve and D a finite collection of points on C.

Proposition 12.5 asserts that all elements of $\mathcal{P}^1$ are of curve type.

Corollary 12.7. *All elements of $\mathcal{P}^1$ are $\mathbb{Z}$-linear combinations of integrals of the form*

$$\int_\gamma \omega,$$

where ω is a regular algebraic 1-form on a smooth affine curve C over $\overline{\mathbb{Q}}$, and γ is a differentiable path on $C(\mathbb{C})$ which is either closed or has endpoints defined over $\overline{\mathbb{Q}}$.

This is a special case of the identification of normal crossings periods and periods of algebraic varieties; see [HMS17, Theorem 11.4.2]. The case $i = 1$ is easier and we give the proof explicitly.

Proof Given Proposition 12.5, this is a statement about the explicit description of the relative singular and de Rham cohomologies. Let C be a smooth affine curve and $Y \subset C$ a finite set of $\overline{\mathbb{Q}}$-points. We made the period computation explicit in Section 3.3.1. All period numbers are of the form

$$((\omega, \alpha), \sigma) = \int_\sigma \omega - \alpha(\partial\sigma)$$

for an algebraic differential form ω on C, a set-theoretic map $\alpha\colon Y(\overline{\mathbb{Q}}) \to \overline{\mathbb{Q}}$ and a formal linear combination $\sigma = \sum n_i \gamma_i$ of smooth maps $\gamma_i\colon [0,1] \to C^{\mathrm{an}}$ such that $\partial(\sum n_i \gamma_i) = \sum n_i \gamma_i(1) - \sum n_i \gamma_i(0)$ is a divisor on $Y(\overline{\mathbb{Q}})$.

The second term appears only for non-closed paths. It is an algebraic number and can be expressed as a period integral of $\mathbb{A}^1$, namely $\int_{[0,1]} \alpha(\partial\sigma)dt$. We conclude that every element of $\mathcal{P}^1$ is a $\mathbb{Z}$-linear combination of explicit integrals as stated.   □

Conversely, the following proposition shows that all periods of curves are in $\mathcal{P}^1$.

Proposition 12.8. *Let C be a curve over $\overline{\mathbb{Q}}$ and $Y \subset C$ a finite set of $\overline{\mathbb{Q}}$-valued points. Then*

$$\mathcal{P}\langle H^*(C, Y)\rangle \subset \mathcal{P}^1.$$

Proof Consider $H^i(C, Y)$ for $0 \le i \le 2$. The assertion holds for $i = 1$ by definition and was shown in Lemma 12.4 for $i = 0$. In the case $i = 2$, dimension reasons show that $H^2(C, Y) \cong H^2(C)$. We replace C by its normalisation. By the blow-up sequence formulated in Proposition 3.12, this does not change $H^2(C)$. Without loss of generality, C is connected. If C is affine, then $H^2(C) = 0$ and $H^2(C) \cong H^2(\mathbb{P}^1)$ if C is projective. From the long exact Mayer–Vietoris sequence

$$H^1(\mathbb{A}^1) \oplus H^1(\mathbb{P}^1 \setminus \{0\}) \to H^1(\mathbb{G}_m) \to H^2(\mathbb{P}^1) \to H^2(\mathbb{A}^1) \oplus H^2(\mathbb{P}^1 \setminus \{0\})$$

for the cover of $\mathbb{P}^1$ by $\mathbb{A}^1$ and $\mathbb{P}^1 \setminus \{\infty\}$ and the vanishing of cohomology of affine spaces, we deduce an isomorphism $H^2(\mathbb{P}^1) \cong H^1(\mathbb{G}_m)$. This shows that the periods agree, which means that the periods of $H^2(\mathbb{P}^1)$ are also in $\mathcal{P}^1$. $\quad\square$

12.3 Comparison with Periods of 1-Motives

From a conceptual point of view, it is also important to describe 1-periods in terms of Jacobians of curves.

Let C be a smooth curve over $\overline{\mathbb{Q}}$, and $D \subset C$ a finite set of $\overline{\mathbb{Q}}$-points. Let $J(C)$ be its generalised Jacobian (see Section 4.5) and

$$\mathbb{Z}[D]^0 = \left\{ f\colon D \to \mathbb{Z} \,\Big|\, \sum_{P \in D} f(P) = 0 \right\}$$

the set of divisors of degree 0 supported in D. We consider the 1-motive

$$M = [\mathbb{Z}[D]^0 \to J(C)].$$

Lemma 12.9. *In this situation, we have*

$$\mathcal{P}(H^1(C, D)) = \mathcal{P}(H^1(J(C), D)) = \mathcal{P}(M).$$

Proof We write D as $D = \{P_0, \ldots, P_r\}$. The easy case when $D = \varnothing$ is left to the reader. The point P_0 is used for the definition of the inclusion $C \to J(C)$ which induces by functoriality a morphism in $(\overline{\mathbb{Q}}, \mathbb{Q})$-Vect

$$H^1(J(C), D) \to H^1(C, D).$$

We apply the long exact cohomology sequence for relative cohomology

$$
\begin{array}{ccccccccc}
H^0(J(C)) & \longrightarrow & H^0(D) & \longrightarrow & H^1(J(C), D) & \longrightarrow & H^1(J(C)) & \longrightarrow & 0 \\
\cong \downarrow & & \| & & \downarrow & & \downarrow & & \\
H^0(C) & \longrightarrow & H^0(D) & \longrightarrow & H^1(C, D) & \longrightarrow & H^1(C) & \longrightarrow & 0
\end{array}
$$

to both terms.

By Theorem 4.23, the induced natural map $H^1(J(C)) \to H^1(C)$ is an isomorphism. According to the Five-Lemma, the same is true for $H^1(J(C), D) \to H^1(C, D)$ and so their periods agree.

In our second step we apply Proposition C.2 to the 1-motive $M = [\mathbb{Z}[D]^0 \to J(C)]$. Note that $e_i = P_i - P_0$ for $i = 1, \dots, r$ is a basis of $\mathbb{Z}[D]^0$, and the natural map $D \to J(C)$ maps P_0 to 0 and all other P_i to the corresponding e_i. By Proposition C.2, this induces an isomorphism

$$V(M)^\vee = H^1(J(C), D), \tag{12.1}$$

and Lemma 7.13 implies

$$\mathcal{P}(M) = \mathcal{P}(H^1(J(C), D)). \qquad \square$$

Having now identified periods of curves with periods of semi-abelian varieties, we can make the step to 1-motives.

Proposition 12.10. *A complex number is a period of some* 1*-motive if and only if it is in* $\mathcal{P}^1$. *In other words,*

$$\mathcal{P}^1 = \mathcal{P}(1\text{-Mot}_{\overline{\mathbb{Q}}}).$$

Proof Let α be in $\mathcal{P}^1$. Proposition 12.5 tells us that it is of curve type, and by Lemma 12.9 it is the period of a 1-motive.

For the converse, let $M = [L \to G]$ be a 1-motive. Up to isogeny, we can split L as $L = L_1 \oplus L_2$ with $L_1 \xrightarrow{0} G$ and $L_2 \hookrightarrow G$. This gives

$$M \cong [L_1 \to 0] \oplus [L_2 \hookrightarrow G],$$

and it therefore suffices to consider the two special cases $G = 0$ or when the structure map $L \to G$ is injective. For $M = [\mathbb{Z} \to 0]$, Proposition C.2 gives $V(M)^\vee \cong H^0(\mathrm{Spec}(\overline{\mathbb{Q}}))$. In fact, the space of periods is simply $\overline{\mathbb{Q}}$ in this case. For $M = [\mathbb{Z}^r \hookrightarrow G]$ we have by Proposition C.2 that $V(M)^\vee \cong H^1(G, Z)$ with $Z = \{0, P_1, \dots, P_r\}$, where P_i is the image of the ith standard basis vector of $\mathbb{Z}^r$. Their periods are in $\mathcal{P}^1$. $\qquad \square$

Summary 12.11. *There are three different definitions of what a* 1*-period might be:*

1. *the period of some* $H^1(X, Y)$ *(cohomological degree* 1*),*
2. *the period of a curve relative to some points (dimension* 1*),*
3. *the period of a Deligne* 1*-motive.*

Our discussion shows that these three notions agree.

12.4 The Motivic Point of View

So far, we have discussed periods of Deligne's category of 1-motives only. There are two other categories of mixed motives, due to Voevodsky and Nori, respectively. In both cases, periods can be defined. The purpose of this section is to compare the sets of numbers that we obtain.

The theories are technically very demanding. It would go too far to present them in detail here. In Appendices A and B the interested reader can find a more complete survey.

The motivic picture gives a lot more structure to the situation. It also shows that most results comparing cohomological 1-periods to other notions of 1-periods can easily be deduced from the literature. The results will be used only in Chapter 13.

Let k be a field with a fixed embedding into $\mathbb{C}$. We denote by $\mathrm{DM}_{\mathrm{gm}}^{\mathrm{eff}}(k, \mathbb{Q})$ Voevodsky's triangulated category of effective geometric motives; see Appendix B. The category comes with a functor which attaches to every smooth k-variety X its *motive* $M(X)$. Let $d_1\mathrm{DM}_{\mathrm{gm}}^{\mathrm{eff}}(k, \mathbb{Q})$ be the full thick sub-category generated by the motives of the form $M(X)$ for X a smooth variety of dimension at most 1. There is a natural equivalence of triangulated categories,

$$D^b(1\text{-}\mathrm{Mot}_k) \to d_1\mathrm{DM}_{\mathrm{gm}}(k, \mathbb{Q}),$$

from the bounded derived category of the abelian category 1-Mot_k to the traingulated category $d_1\mathrm{DM}_{\mathrm{gm}}(k, \mathbb{Q})$. See Theorem B.5 for more details.

Next, we turn to Nori's abelian category $\mathcal{MM}_{\mathrm{Nori}}^{\mathrm{eff}}(k, \mathbb{Q})$ of effective motives. It is universal for all cohomological functors compatible with rational singular cohomology. In Appendix A a very brief introduction to Nori's theory is given.

Let $d_1\mathcal{MM}_{\mathrm{Nori}}^{\mathrm{eff}}(k, \mathbb{Q})$ be the smallest full subcategory containing $H_{\mathrm{Nori}}^i(X, Y)$ for $Y \subset X$ with $i \leq 1$ and which is closed under subquotients. Again, by the work of Ayoub and Barbieri-Viale (see Theorem A.7), there is an anti-equivalence

$$1\text{-}\mathrm{Mot}_k \to d_1\mathcal{MM}_{\mathrm{Nori}}^{\mathrm{eff}}(k, \mathbb{Q}).$$

Moreover, the abelian category $d_1\mathcal{MM}_{\mathrm{Nori}}^{\mathrm{eff}}(k, \mathbb{Q})$ has an explicit description as the diagram category (in the sense of Nori) of the category of pairs (C, Y), where C is a smooth curve and $Y \subset C(k)$ a finite subset.

By Theorem B.6, both categories are linked by a triangulated realisation functor

$$\mathrm{DM}_{\mathrm{gm}}^{\mathrm{eff}}(k, \mathbb{Q}) \to D^b(\mathcal{MM}_{\mathrm{Nori}}^{\mathrm{eff}}(k, \mathbb{Q}))$$

compatible with their singular realisations into the derived category of $\mathbb{Q}$-vector spaces. By Proposition B.7, it maps the subcategory $d_1\mathrm{DM}_{\mathrm{gm}}(k, \mathbb{Q})$ to $D^b(d_1\mathcal{MM}_{\mathrm{Nori}}(k, \mathbb{Q}))$.

The universal property of Nori motives implies the existence of functors

$$\mathcal{MM}_{\text{Nori}}^{\text{eff}}(k, \mathbb{Q}) \hookrightarrow \text{MHS}_k \hookrightarrow (\mathbb{Q}, k)\text{-Vect};$$

see Theorem A.2. The first functor associates to a Nori motive a mixed Hodge structure. By forgetting the filtrations, we obtain an object of $(\mathbb{Q}, k)$-Vect. This allows us to define periods for the various categories of motives. All this is summed up in one diagram:

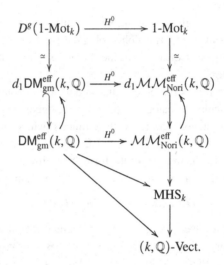

By Proposition C.3, the composition in the right column is the functor $V \colon 1\text{-Mot}_k \to (\mathbb{Q}, k)$-Vect of Section 8.2 composed with the external duality functor of Definition 7.12.

Corollary 12.12. *The sets of periods of the categories* 1-Mot_k, $d_1 \text{DM}_{\text{gm}}^{\text{eff}}(k, \mathbb{Q})$ *and* $d_1 \mathcal{MM}_{\text{Nori}}^{\text{eff}}(k, \mathbb{Q})$ *agree and are equal to* $\mathcal{P}^1$. *In particular, both Proposition 12.5 and Proposition 12.10 hold true.*

Proof Theorems B.5 and B.6 combined with Theorem A.7 provide the equivalences of categories

$$D^b(1\text{-Mot}_k) \to d_1 \text{DM}_{\text{gm}}^{\text{eff}}(k, \mathbb{Q}) \to D^b\big(d_1 \mathcal{MM}_{\text{Nori}}^{\text{eff}}(k, \mathbb{Q})\big).$$

Proposition C.3 asserts that Deligne's construction of the realisation of a 1-motive agrees (up to duality) with the one obtained via the identification with the category $d_1 \mathcal{MM}_{\text{Nori}}^{\text{eff}}(k, \mathbb{Q})$. As a consequence, the three categories have identical sets of periods. Finally, the periods of $D^b(1\text{-Mot}_k)$ coincide with the periods of 1-Mot_k by definition.

Also by definition, $\mathcal{P}^1 \subset \mathcal{P}(d_1 \mathcal{M}\mathcal{M}^{\text{eff}}_{\text{Nori}}(k, \mathbb{Q}))$. The explicit description of $d_1 \mathcal{M}\mathcal{M}^{\text{eff}}_{\text{Nori}}(k, \mathbb{Q})$ in Theorem A.7 yields the converse inclusion and even the more restrictive description of Proposition 12.5.

As a by-product, we get the equality $\mathcal{P}^1 = \mathcal{P}(1\text{-Mot}_k)$, reproving Proposition 12.10. $\qquad\qquad\square$

13

Relations Between Periods

In the previous chapter, we established different descriptions for the space of 1-periods. We now turn to their relations.

13.1 Kontsevich's Period Conjecture

There is a short list of obvious relations.

(A) *Bilinearity:* Let X be a variety over $\overline{\mathbb{Q}}$, $Y \subset X$ a closed subvariety and $i \in \mathbb{N}_0$. For all $\sigma_1, \sigma_2 \in H_i^{\text{sing}}(X, Y; \mathbb{Q})$, $\omega_1, \omega_2 \in H_{\text{dR}}^i(X, Y)^\vee$, $\mu_1, \mu_2 \in \mathbb{Q}$, $\lambda_1, \lambda_2 \in \overline{\mathbb{Q}}$, we have

$$\int_{\mu_1\sigma_1 + \mu_2\sigma_2} (\lambda_1\omega_1 + \lambda_2\omega_2) = \sum_{i,j=1,2} \mu_i\lambda_j \int_{\sigma_i} \omega_j.$$

(B) *Functoriality:* Let $f \colon (X, Y) \to (X', Y')$ be a morphism of pairs of $\overline{\mathbb{Q}}$-varieties and $i \in \mathbb{N}_0$. For all $\sigma \in H_i^{\text{sing}}(X, Y; \mathbb{Q})$ and $\omega' \in H_{\text{dR}}^i(X', Y')$, we have

$$\int_\sigma f^*\omega' = \int_{f_*\sigma} \omega'.$$

(C) *Boundary maps:* Let $X \supset Y \supset Z$ be subvarieties and $i \in \mathbb{N}_0$. For all $\sigma \in H_{i+1}^{\text{sing}}(X, Y; \mathbb{Q})$ and $\omega \in H_{\text{dR}}^i(Y, Z)$, we have

$$\int_{\partial\sigma} \omega = \int_\sigma \partial\omega,$$

where ∂ denotes the boundary maps $H_{\text{dR}}^i(Y, Z) \to H_{\text{dR}}^{i+1}(X, Y)$ on de Rham cohomology and $H_{i+1}^{\text{sing}}(X, Y; \mathbb{Q}) \to H_{\text{sing}}^i(Y, Z; \mathbb{Q})$ on singular homology, respectively.

To state the Period Conjecture we recall from Definition 12.2 the set of i-periods $\mathcal{P}^i \subset \mathbb{C}$. Let $\mathcal{P}^{\text{eff}} = \bigcup_{i=0}^{\infty} \mathcal{P}^i$ be the set of *effective cohomological periods* and $\mathcal{P} = \mathcal{P}^{\text{eff}}[\pi^{-1}]$ the *period algebra*.

Conjecture 13.1 (Period Conjecture, Kontsevich [Kon99]). *All $\overline{\mathbb{Q}}$-linear relations between elements of $\mathcal{P}$ are induced by the above relations.*

Remarks 13.2. 1. In the abstract formalism of Chapter 7 this is the Period Conjecture for the diagram Pairs^{eff} for $k = \overline{\mathbb{Q}}$; see Definition A.1.
2. A close look shows that Conjecture 13.1 is not identical to the conjecture originally formulated by Kontsevich in [Kon99]. He considered only smooth varieties X and divisors with normal crossings Y. We refer the reader to the discussion in [HMS17, Remark 13.1.8] for the precise connection. The version above implies that $\mathrm{Spec}(\mathcal{P})$ is a torsor under the motivic Galois group of $\overline{\mathbb{Q}}$, a result due to Nori. It was first formulated in [Kon99, Theorem 6]. A complete proof can be found in [HMS17, Theorem 13.1.4].
3. By the Künneth formula, products of periods are in fact periods of products of varieties. Hence the above conjecture also says something about *algebraic* relations between periods and, indeed, it is equivalent to a Grothendieck style version of the Period Conjecture. For a complete discussion, see [HMS17, Section 13.2]. We do not deal with the latter because we are interested in the set $\mathcal{P}^1$, which is *not* closed under multiplication.

Theorem 13.3 (Period Conjecture for $\mathcal{P}^1$). *The Period Conjecture is true for the subset $\mathcal{P}^1$. More explicitly, the following equivalent statements hold.*

1. *All relations between periods of 1-motives are induced by bilinearity and functoriality of 1-motives.*
2. *All relations between periods of curve type are induced by bilinearity and functoriality of pairs $(C, D) \to (C', D')$, with C, C' smooth affine curves and D, D' finite sets of points of C and C', respectively.*
3. *All relations between periods in cohomological degree at most 1 are induced by the relations (A), (B) and (C).*

Remark 13.4. This theorem does not mention Nori motives. In contrast to Chapter 12, we have not been able to eliminate them from the proofs, at least not without disproportionate effort.

Motivic proof of Theorem 13.3. Assertion (1) is precisely the statement of Theorem 9.10. Hence it remains to show the equivalence with the others. Here we rely significantly on the results of Appendix A.

By Theorem A.7, the category 1-Mot$_{\overline{\mathbb{Q}}}$ is equivalent to $d_1 \mathcal{MM}_{\text{Nori}}^{\text{eff}}(\overline{\mathbb{Q}}, \mathbb{Q})$. This category has a description as the diagram category of the diagram of pairs

(C, D) with C a smooth affine curve and D a finite set of points on D. By the general results of [HMS17, Theorem 8.4.22], this implies that all relations are induced by bilinearity and functoriality for the edges of the diagram, i.e. functoriality for pairs. This is the implication from assertion (1) to assertion (2).

The proof of Proposition 12.5 shows that all elements in $\mathcal{P}^1$ can be related to periods of curve type using only the operations (A), (B), (C). Hence assertion (2) implies assertion (3).

In order to show the implication from assertion (3) to assertion (1), we apply Theorem A.7 and replace 1-Mot$_{\overline{\mathbb{Q}}}$ by the equivalent $d_1 \mathcal{MM}_{\text{Nori}}^{\text{eff}}(\overline{\mathbb{Q}}, \mathbb{Q})$. Every object M in the latter category is a subquotient of an object of the form $H^i(X, Y)$ for $i \leq 1$. Theorem A.7 even shows that it suffices to take $i = 1$ and X a curve. The functoriality relation for periods of Nori motives identifies the periods of M with the periods of $H^i(X, Y)$ for $i \leq 1$. The relations (B) and (C) are special cases of the functoriality relation for Nori-1-motives. This completes the proof. $\qquad\square$

We come back to the category of Nori-1-motives and its realisations as discussed in Section 12.4.

Theorem 13.5 (Fullness). *The three natural functors f_1, f_2, f_3 on* 1-Mot$_{\overline{\mathbb{Q}}}$,

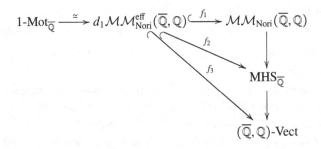

are fully faithful with image closed under subquotients.

Proof It suffices to consider the total functor f_3. Composition of f_3 with the anti-equivalence with 1-Mot$_{\overline{\mathbb{Q}}}$ is simply $\cdot \circ V$. Therefore it suffices to consider $V\colon$ 1-Mot$_{\overline{\mathbb{Q}}} \to (\mathbb{Q}, \overline{\mathbb{Q}})$-Vect. This functor is fully faithful by Theorem 9.14. $\quad\square$

Remark 13.6. We gave a direct proof for MHS$_{\overline{\mathbb{Q}}}$ earlier; see Proposition 8.17. Both arguments rely on the Analytic Subgroup Theorem applied to the graph of a morphism, but applied in a different way.

13.2 The Case of Curves

We turn now to the case of curves, which is of particular interest. We begin with some motivating discussion with historical background. Then we turn to the Period Conjecture for curves. One of the highlights is a precise criterion for a sum of periods of a single differential to be algebraic.

13.2.1 Motivating Examples

In his book on transcendental numbers [Sie49], Siegel mentioned several problems which were not accessible at the time. He wrote (see [Sie49, p. 97]):

All our transcendence proofs made essential use of the fact that the problem can be reduced to the proof of a property of entire functions. This is the reason why the known methods do not work for elliptic integrals of the third kind and not even for integrals of the third kind in the still simpler fact of curves of genus 0. For instance, it is not known whether the number

$$\int_0^1 \frac{dx}{1+x^3} = \frac{1}{3}\left(\log 2 + \frac{\pi}{\sqrt{3}}\right)$$

is irrational.

Integrals of this form along not necessarily closed paths are what we call *incomplete periods of the third kind* on $\mathbb{P}^1$. As it turned out, such integrals are not only irrational but also transcendental, as follows from Baker's work on linear forms in logarithms. Indeed, one deduces from the inhomogeneous case of Baker's Theorem about linear forms in logarithms of algebraic numbers that the numbers $1, \log 2$ and $\pi = -i\log(-1)$ are linearly independent over $\overline{\mathbb{Q}}$. Strictly speaking, this is not a transcendence result but a result on linear independence of incomplete periods of the third kind in the case of a curve of genus 0. However, the transcendence of $\log 2$ and $\pi = -i\log(-1)$ is an immediate consequence.

A. van der Poorten (see [VdP71]) considered a more general complete period and also an incomplete period of the third kind on a curve of genus 0. In this case a differential ξ of the third kind takes the form

$$\xi = \frac{P(x)}{Q(x)}dx,$$

where $P(x)$ and $Q(x)$ are polynomials. He considers a path $\gamma: [0,1] \to \mathbb{P}^1$ along which the differential form is defined and which satisfies $\gamma(0), \gamma(1) \in \mathbb{P}^1(\overline{\mathbb{Q}})$. We write $\alpha_1, \ldots, \alpha_n$ for the zeroes of Q and $r_1, \ldots, r_n$ for the residues at the poles of the differential form ξ. Then van der Poorten deduces again from the inhomogeneous version of Baker's Theorem on linear forms in logarithms, that $\int_\gamma \xi$ is algebraic if and only

$$\int_\gamma \left(\sum_{k=1}^{n} \frac{r_k}{x - \alpha_k} \right) dx = 0.$$

This follows from taking the partial fraction decomposition

$$\frac{P(x)}{Q(x)} dx = dF(x) + \sum_{k=1}^{n} \frac{r_k}{x - \alpha_k} dx$$

and integrating along γ.

In Theorem 13.9 below we will give a generalisation of van der Poorten's result to curves of any genus. In particular van der Poorten's Theorem is a special case of our result. Furthermore, in the case of positive genus, this includes abelian integrals of the third kind and proves transcendence of complete and incomplete periods.

An even older example was pointed out by Arnol'd in [Arn90]. He makes reference to a letter from Leibniz to Huygens, dated 10/20 April 1691. In this letter Leibniz formulated the problem of transcendence of the areas of segments cut off from an algebraic curve, defined by an equation with rational coefficients, by straight lines with algebraic coefficients (see [Arn90, p. 93n]). In [Arn90, p. 105] Arnol'd reformulated this problem, turning it into modern language: an abelian integral along an algebraic curve with rational (algebraic) coefficients taken between limits which are rational (algebraic) numbers is generally a transcendental number. Again Theorem 13.9 below gives the solution to Leibniz's problem. We refer the reader to [Wüs12] for a more detailed discussion of the example.

As Arnol'd pointed out, the problem is also very interesting from a historical point of view, in so far as it was previously believed that transcendence theory developed in the nineteenth century with Liouville, Hermite and others. The document of Leibniz, however, shows that the concept of transcendence of numbers was already present in the seventeenth century.

13.2.2 The Period Conjecture for Curves

Theorem 13.7. *Let C be a smooth curve over $\overline{\mathbb{Q}}$ with generalised Jacobian $J(C)$ and $D \subset C$ a finite set of $\overline{\mathbb{Q}}$-points with group of divisors of degree 0 supported in D denoted $L = \mathbb{Z}[D]^0$. Then all relations between periods of $H^1(C, D)$ are induced by bilinearity and morphisms between subquotients of sums of the 1-motive $[L \to J(C)]$.*

Proof By Lemma 12.9, the periods of $H^1(C, D)$ agree with the periods of the 1-motive as described in the theorem. We then apply Theorem 9.10. □

Remark 13.8. Note that this version of the Period Conjecture does not rely on the higher theory of geometric or Nori motives – only 1-motives are used. This version is actually a lot more useful in computations.

We now turn from relations between periods to the question of transcendence of periods of differential forms. It suffices to consider the case of a smooth projective curve C. With a rational function $f \in \overline{\mathbb{Q}}(C)^*$ we associate a meromorphic differential form $\omega = df$. We also choose a path γ in C^{an} which avoids the singularities of f and has starting point and endpoint in $C(\overline{\mathbb{Q}})$. Then

$$\int_\gamma \omega = f(\gamma(1)) - f(\gamma(0)) \in \overline{\mathbb{Q}}.$$

This is essentially the only way to produce algebraic periods from meromorphic differential forms, as the following theorem shows.

Theorem 13.9 (Transcendence of periods). *Let C be a smooth projective curve over $\overline{\mathbb{Q}}$ and ω a meromorphic differential form defined over $\overline{\mathbb{Q}}$. Let $\sigma = \sum_{i=1}^n a_i \gamma_i$, where $\gamma_i \colon [0,1] \to C$ for $i = 1, \ldots, n$ are differentiable paths avoiding the poles of ω and $a_i \in \mathbb{Z}$. We assume that $\partial\sigma$ has support in $C(\overline{\mathbb{Q}})$. In this situation the period*

$$\alpha = \int_\sigma \omega$$

is algebraic if and only if ω is the sum of an exact form with no extra poles and a form with vanishing period.

Remark 13.10. The theorem includes famous cases like the transcendence of π, $\log\alpha$ (for α algebraic) and periods and quasi-periods of elliptic curves. Forms of the first, second and third kind are allowed.

Proof Let $C^\circ \subset C$ be an affine curve such that ω is holomorphic on C°. Let $D \subset C^\circ$ be the set of starting points and endpoints of the paths $\gamma_1, \ldots, \gamma_n$. Then α can be considered as a period of $H^1(C^\circ, D)$. We introduce the generalised Jacobian $G = J(C^\circ)$ and fix an embedding $C^\circ \to G$. This translates α into a period of the 1-motive $M = [\mathbb{Z}[D]^0 \to G]$ by viewing $[\omega] \in H^1_{\mathrm{dR}}(C^\circ, D)$ as an element of $V^\vee_{\mathrm{dR}}(M)$ and $[\sigma] \in H^{\mathrm{sing}}_1(C^\circ, D; \mathbb{Q})$ as an element of $V_{\mathrm{sing}}(M)$. By Theorem 9.11, the algebraicity of the period implies that ω can be written as $\phi + \psi$ such that $\int_\sigma \psi = 0$ and the image of ϕ in $V_{\mathrm{dR}}(G)^\vee = H^1_{\mathrm{dR}}(G) \cong H^1_{\mathrm{dR}}(C^\circ)$ is zero. As C° is affine, this means that the differential form $\phi \in \Omega^1(C^\circ)$ is exact with poles only in $C \smallsetminus C^\circ$.

This completes the proof except when $\omega \in \Omega^1(C)$. It remains to show that ϕ is not only exact but has no poles. This requires an extra argument, which we give in Proposition 13.11 below. □

Proposition 13.11. *Let C be a smooth projective curve over* $\overline{\mathbb{Q}}$, $\omega \in \Omega^1(C)$ *and* σ *a linear combination of paths with endpoints in* $C(\overline{\mathbb{Q}})$. *If* $\int_\sigma \omega$ *is algebraic, then it is zero.*

Proof Let $M = [\mathbb{Z}[D]^0 \to J(C)]$ be as in the proof of Theorem 13.9. We consider $\sigma \in V_{\mathrm{sing}}(M)$ as an element of $\mathrm{Lie}(M^\natural)$ via the inclusion $V_{\mathrm{sing}}(M) \subset \mathrm{Lie}(M^\natural)$. By construction $M^\natural$ is a vector extension of $J(C)$ and by assumption ω is in the image of $\mathrm{coLie}(J(C)) \to \mathrm{coLie}(M^\natural)$. Hence the period α depends only on the image $\bar{\sigma}$ of σ in $\mathrm{Lie}(J(C))_{\mathbb{C}}$. We now consider the connected algebraic group $G = J(C) \times \mathbb{G}_a$ and apply the Analytic Subgroup Theorem 6.2 to the point $u = (\bar{\sigma}, 1) \in \mathrm{Lie}(G)^{\mathrm{an}}$ and to $(\omega, \alpha dt) \in \mathrm{coLie}(G)$. The theorem gives a short exact sequence

$$0 \to H \to G \xrightarrow{\pi} G/H \to 0$$

such that u is in the image of $\mathrm{Lie}(H)^{\mathrm{an}}$ and $(\omega, \alpha dt)$ in the image of $\mathrm{coLie}(G/H)$ under π^*. As G is the product of an abelian variety and $\mathbb{G}_a$, there are only two possibilities for H: it is either an abelian subvariety of $J(C)$ or a product of an abelian subvariety B with $\mathbb{G}_a$. In the first case, the inclusion $H \to G$ factors via $J(C)$. This contradict the shape of u. In the second case, $G/H = J(C)/B$ is an abelian variety. The subgroup $\mathbb{G}_a$ is contained in the kernel of π and so $\pi^*\mathrm{coLie}(G/H) \subset \mathrm{coLie}(J(C)) \times 0$, i.e. $\alpha dt = 0$. This gives $\alpha = 0$ as we aimed to prove. □

Remark 13.12. This proof goes back to the original formulation of the Analytic Subgroup Theorem. It is not enough to apply the transcendence criterion in Theorem 9.11. The additional input here is the Hodge filtration on $V_{\mathrm{dR}}(M)$.

Corollary 13.13. *Let* C, ω *and* σ *be as in Theorem 13.9. If in addition* $\omega \in \Omega^1(C)$ *or if the divisor* $\partial\sigma = \sum_{i=1}^n a_i\gamma_i(1) - \sum_{i=1}^n a_i\gamma_i(0)$ *vanishes, then the period* α *is either transcendental or zero.*

Proof Suppose that $\alpha = \int_\sigma \omega$ is algebraic. By the theorem, there is a decomposition $\omega = df + \phi$ such that $\int_\sigma \phi = 0$. If the boundary divisor vanishes, then

$$\int_\sigma df = \int_{\partial\sigma} f = 0$$

for all f. If $\omega \in \Omega^1(C)$, then f is in $\mathcal{O}(C)$, which gives $df = 0$. In both cases, $\alpha = \int_\sigma \phi = 0$. □

Masser pointed out to us that most of Theorem 13.9 has an elementary reduction to the case of closed cycles announced in [Wüs87]. We explain a variant of his argument.

Proof Let σ be as in the theorem such that $\partial\sigma \neq 0$. We write $\partial\sigma = \sum_{i=1}^{m} b_i P_i$ with $P_i \in C(\overline{\mathbb{Q}})$ and non-vanishing b_i. By assumption, $m \geq 1$. Let Q be in the polar locus of ω. (If $\omega \in \Omega^1(C)$ pick any $Q \in C(\overline{\mathbb{Q}})$ not on any of the $\gamma_i([0,1])$.)

We consider the divisors $D_N = NQ - P_2 - \cdots - P_m$ and $D_N' = D_N - P_1$. For big enough N, Riemann–Roch gives $l(D_N) = l(D_N') - 1$. Let $f \in L(D_N') \setminus L(D_N)$. This is a rational function with a pole only in Q and a zero in $P_2, \ldots, P_m$, but not in P_1. As a consequence, we obtain

$$\int_\sigma df = \sum_{i=1}^{m} b_i f(P_i) = b_1 f(P_1) \neq 0.$$

The function f can be normalised such that the value of the integral is 1. If $\alpha = \int_\sigma \omega$ is algebraic, then it is equal to $\int_\sigma d(\alpha f)$. Introduce $\phi = \omega - d(\alpha f)$. By construction, $\int_\sigma \phi = 0$, as we wanted to show.

As in the original proof of Theorem 13.9, the argument does not allow us to control the poles of df in the special case where ω is holomorphic. We deduced this case directly from the Analytic Subgroup Theorem in Proposition 13.11. □

Remark 13.14. Qualitatively, we are in a situation very similar to the case of closed cycles: understanding algebraicity of periods requires understanding vanishing of periods.

We come back to the case where $\sigma = \gamma$ is a single non-closed path later in Corollary 14.22 under the simplifying assumption that $J(C)$ is simple.

14

Vanishing of Periods of Curves

In this chapter, we translate the results obtained so far to the classical language and turn to the subtle question of when period integrals on curves vanish. In this chapter, let C be a smooth projective curve over $\overline{\mathbb{Q}}$ and $\omega \in \Omega^1_{\overline{\mathbb{Q}}(C)}$ a rational algebraic differential form on C.

14.1 Classical Periods

We come back to the classical terminology.

Definitions 14.1. We say that ω is

1. *exact* if it is of the form $\omega = df$ for some $f \in \overline{\mathbb{Q}}(C)^*$;
2. *of the first kind* if it does not have poles, i.e. $f \in \Omega^1(C)$;
3. *of the second kind* if the residues of ω are zero;
4. *of the third kind* if it has at most simple poles.

This terminology goes back to the early nineteenth century, when Legendre studied elliptic integrals. We refer the reader to [Wel17, Chapter 7] for historical comments.

Following these definitions, exact forms are of the second kind, and differential forms of the first kind are also of the second and third kind.

Lemma 14.2. *Every differential form can be written in the form*

$$\omega = \omega_2 + \omega_3,$$

with ω_2 of the second kind and ω_3 of the third kind.

Proof Let ω be an arbitrary meromorphic differential form. The sum of its residues is 0. For example, by [GH78, lemma p. 233], there is a form ω_3 of the third kind with the same poles and residues as ω. Then $\omega_2 = \omega - \omega_3$ has the desired property. □

Definitions 14.3. A complex number is called a *classical period* if it is of the form

$$\int_{\sigma} \omega,$$

where $\sigma = \sum_{i=1}^{n} a_i \gamma_i$ is a formal $\mathbb{Z}$-linear combination of C^{∞}-paths avoiding the singularities of ω and with endpoints in $C(\overline{\mathbb{Q}})$. We say that it is

1. *simple* if $n = 1$ (integral over a single path);
2. *complete* if all γ_i are closed;
3. of *the first, second or third kind* in the case that ω is of the first, second or third kind, respectively.

Note that the classical periods are algebraic for exact forms. A cycle $\sigma = \sum_{i=1}^{n} a_i \gamma_i$ with closed γ_i can be seen in the abelianisation of the fundamental group. In consequence, we may replace σ by a homologous closed path defining the same period number. Accordingly all complete periods are simple. However, not all incomplete periods are simple.

We get back the same periods that we considered earlier.

Lemma 14.4. *The set of all classical periods agrees with the set of cohomological periods* $\mathcal{P}^1$.

Proof By Corollary 12.7, all elements of $\mathcal{P}^1$ are classical periods. Conversely, let C° be the complement of the set of poles of ω and $D \subset C^{\circ}(\overline{\mathbb{Q}})$ the finite set of endpoints of the γ_i. Then ω defines a class $[\omega] \in H^1_{dR}(C^{\circ}, D)$ and σ defines a class $[\sigma] \in H^{\mathrm{sing}}_1(C^{\circ,\mathrm{an}}, D; \mathbb{Z})$. The integral computes the period pairing

$$\int_{\sigma} \omega = \langle [\omega], [\sigma] \rangle$$

and makes our classical period a cohomological period. □

Our next aim is to find necessary and sufficient conditions for a differential form ω and a cycle σ to satisfy

$$\int_{\sigma} \omega = 0.$$

Here are some examples:

Examples 14.5. 1. Let γ be contractible in the complement of the set of poles of ω. Then the period vanishes by the Monodromy Theorem.
2. Let ω be of the third kind with poles in $Q_1, \ldots, Q_m$. For every $i = 1, \ldots, m$ let γ_i be the positively oriented boundary of a small disc in C^{an} centred at Q_i. Then

$$\int_{\Sigma \gamma_i} \omega = \sum_{i=1}^{m} \mathrm{res}_{Q_i} \omega = 0$$

by the Residue Theorem.

3. To give an example for vanishing of periods, we take the elliptic curve E given by $y^2 = x^3 + 1$ and define a curve C by $y^2 = x^6 + 1$. Sending (x, y) to (x^2, y) defines a morphism $\pi\colon C \to E$ of degree 2. The genus of E is 1 and the genus of C is 2. The morphism π induces a homomorphism

$$\pi_*\colon H_1(C, \mathbb{Z}) \to H_1(E, \mathbb{Z}).$$

We take $\omega \in H^0(E, \Omega^1(E))$ and $\gamma \in \ker \pi_*$ and obtain

$$\int_\gamma \pi^* \omega = \int_{\pi_* \gamma} \omega = \int_0 \omega = 0.$$

This shows that there are non-trivial examples for vanishing.

We may ask how complete this list is. This question will be addressed by first moving from paths and differential forms to homology and de Rham cohomology and then to 1-motives.

14.2 The Setting

The following notation is fixed for the rest of the chapter.

Notation 14.6.

- C denotes a smooth projective curve over $\overline{\mathbb{Q}}$ with base point $P_0 \in C(\overline{\mathbb{Q}})$.
- $\omega \in \Omega^1_{\overline{\mathbb{Q}}(C)}$ is a meromorphic differential form.
- $S = \{Q_1, \ldots, Q_m\}$ stands for the set of poles of ω with non-trivial residue and $S' = \{R_1, \ldots, R_k\}$ for the set of poles with vanishing residue; without loss of generality, $P_0 \notin S \cup S'$.
- $C^\circ \subset C$ signifies the complement of the set of poles $S = \{Q_1, \ldots, Q_m\}$ of ω.
- $J(C)$ and $J(C^\circ)$ are the (generalised) Jacobians of C and C° (in the sense of Section 4.5) with embeddings $v^\circ\colon C^\circ \to J(C^\circ)$ and $v\colon C \to J(C)$ via $P \mapsto P - P_0$.
- $\sigma = \sum_{i=1}^n a_i \gamma_i$ is a formal $\mathbb{Z}$-linear combination of C^∞-paths $\gamma_i\colon [0, 1] \to C^{\circ, \mathrm{an}} \smallsetminus S'$ with endpoints defined over $\overline{\mathbb{Q}}$.
- Let $D \subset C^\circ \smallsetminus S'$ be a set of points such that the divisor $\partial \sigma$ has support on D. We define $r = |D| - 1$ if $D \neq \emptyset$, and $r = 0$ if $D = \emptyset$.
- $\alpha = \int_\sigma \omega$ is the period of σ and ω.

In Section 5.3 we introduced the map I which assigns to a path, or more generally to a chain in a complex Lie group, an element of the complex Lie algebra. Given a path $\gamma\colon [0, 1] \to C^{\circ, \mathrm{an}}$, we put

$$I^\circ(\gamma) = I(v^\circ \circ \gamma) \in \mathrm{Lie}(J(C^\circ))^{\mathrm{an}}.$$

The operator $l°$ has the property that

$$\exp(l°(\gamma)) = \gamma(1) - \gamma(0) =: P(\gamma) \in J(C°)^{\mathrm{an}},$$

hence should be seen as a choice of logarithm. We extend $l°$ linearly to

$$l°(\sigma) := \sum_{i=1}^{n} a_i l°(\gamma_i).$$

Then

$$\exp(l°(\sigma)) = \sum_{i=1}^{n} a_i P(\gamma_i) =: P(\sigma).$$

Let $l(\sigma)$ be the image of $l°(\sigma)$ in $\mathrm{Lie}(J(C)^{\mathrm{an}})$.

14.2.1 Homological Interpretation

Obviously the chain σ defines an element $[\sigma] \in H_1^{\mathrm{sing}}(C°, D; \mathbb{Z})$. Less obviously, the rational differential form ω defines an element $[\omega]$ of $H_{\mathrm{dR}}^1(C°, D)$; see Section C.1 in Appendix C, and equation (3.1). The argument is well known for $D = \varnothing$. We explain the construction in general.

If $S \cup S' = \varnothing$, then ω is in $\Omega^1(C) \subset H_{\mathrm{dR}}^1(C, D)$ and there is nothing to show. If $S \cup S' \neq \varnothing$, the curve $C° \setminus S'$ is affine and $\omega|_{C° \setminus S'} \in \Omega^1(C° \setminus S')$ defines an element of $H_{\mathrm{dR}}^1(C° \setminus S', D)$. By definition of S', the cohomology class is in the kernel of the residue map $H_{\mathrm{dR}}^1(C° \setminus S', D) \to H^0(S')(-1)$, making it even an element of $H_{\mathrm{dR}}^1(C°, D)$. We carry out the cocycle computation for $[\omega]$ in the representation of Lemma 3.20.

Recall that $S' = \{R_1, \ldots, R_k\}$ and choose $f_i \in \overline{\mathbb{Q}}(C)^*$ such that the principal part of df_i at R_i coincides with the principal part of ω in R_i. Such a function exists because the residue of ω vanishes. We write $\omega_i = \omega - df_i$ and $U_i \subset C$ for the complement of the set of poles of ω_i. By construction, $R_i \in U_i$. We also introduce $U_0 = C° \setminus S'$, $\omega_0 = \omega$, $f_0 = 0$. Then $\mathfrak{U} = \{U_0, \ldots, U_k\}$ is an open cover of $C°$. The differential form ω defines a cocycle in $H_{\mathrm{dR}}^1(C° \setminus S', D; \mathfrak{U})$ given by the tuple $\underline{\omega} = (\omega|_{U_i \setminus S'}, 0, 0)$, cohomologous to the cocycle

$$\underline{\omega} - \partial f = ((\omega - df_i)_i, (-f_j + f_i)_{i,j}, (f_i|_D)_i),$$

where i, j run through 0 to k. It defines a cocycle in $H_{\mathrm{dR}}^1(C°, D; \mathfrak{U})$ as required.

Lemma 14.7. *The class* $[\omega] \in H_{\mathrm{dR}}^1(C°, D)$ *is zero if and only if* ω *can be represented as* $\omega = df$ *for a function* $f \in \overline{\mathbb{Q}}(C)^*$ *which is regular in* D *and vanishes there. Moreover, every element of* $H_{\mathrm{dR}}^1(C°, D)$ *is of the form* $[\omega]$ *such that the set of poles of* ω *with non-trivial residue is contained in* S.

Proof If $[\omega] = 0$, then by Lemma 3.20 the cocycle is of the form

$$\underline{\omega} - \partial f = ((dg_i)_i, (g_j - g_i)_{ij}, (-g_i|_{D_i})_i)$$

for $(g_i)_i \in \prod_i \mathcal{O}(U_i)$; in particular,

$$-f_j + f_i = g_j - g_i \in \mathcal{O}(U_i \cap U_j)$$

for all i, j. This implies that the collection of functions $f_i + g_i \in \mathcal{O}(U_i)$ glues to a global function $F \in \mathcal{O}(C^\circ)$. Moreover,

$$\omega - df_i = dg_i \in \Omega^1(U_i)$$

for all i and hence $dF = \omega$. We then have

$$\underline{\omega} - \partial f \sim (dF|_{U_i}, 0, 0) \sim (0, 0, F|_{D_i}).$$

Note that the g_i and hence also F are unique up to an additive constant. The triviality of $[\omega]$ implies that $F|_D$ is constant. By adjusting the constant, we get $\omega = dF$ with $F|_D = 0$ as claimed.

It remains to prove that every class in $H^1_{\mathrm{dR}}(C^\circ, D)$ can be represented by a differential form. It is well known that every element of $H^1_{\mathrm{dR}}(C)$ is represented by a differential form of the second kind; see [GH78, p. 459]. The sequence

$$H^1_{\mathrm{dR}}(C) \to H^1_{\mathrm{dR}}(C^\circ) \xrightarrow{\mathrm{res}} H^0_{\mathrm{dR}}(S)(-1) \to H^1_{\mathrm{dR}}(C)(-1)$$

for relative cohomology is exact. For every element of $H^0_{\mathrm{dR}}(S) \cong \overline{\mathbb{Q}}^m$ summing to 0 in $H^0_{\mathrm{dR}}(C) \cong \overline{\mathbb{Q}}$, there is by [GH78, lemma p. 233] a form ω_3 of the third kind with these residues.

This shows that given a class $c \in H^1_{\mathrm{dR}}(C^\circ)$, there exists ω_3 of the third kind with the same residues. The difference $c - [\omega_3]$ is in the image of $H^1_{\mathrm{dR}}(C)$ and represented by a differential form of the second kind.

The sequence

$$H^0_{\mathrm{dR}}(D) \to H^1_{\mathrm{dR}}(C^\circ, D) \to H^1_{\mathrm{dR}}(C^\circ)$$

for the relative cohomology is also exact. Elements of $H^0_{\mathrm{dR}}(D)$ are functions $f : D \to \overline{\mathbb{Q}}$. Let $U_0 \subset C^\circ$ be an open affine subset containing D and $U_0, \ldots, U_n$ an affine cover of C°. The image of f in $H^1_{\mathrm{dR}}(C^\circ, D, \mathfrak{U})$ is the cocycle

$$(0, 0, f|_{D_i}).$$

As U_0 is affine, the closed immersion $D \subset U_0$ induces a surjection $\mathcal{O}(U_0) \twoheadrightarrow \mathcal{O}(D)$. We choose a lift $\widetilde{f} \in \mathcal{O}(U_0)$ of f. The class of $d\widetilde{f}$ agrees with the class of $(0, 0, f|_{D_i})$ from above.

Given a class c in $H^1_{\mathrm{dR}}(C^\circ, D)$, we have shown that there is a form ω with correct residues such that the images of c and $[\omega] \in H^1_{\mathrm{dR}}(C^\circ, D)$ in $H^1_{\mathrm{dR}}(C^\circ)$

agree. Their difference is in the image of $H_{\mathrm{dR}}^0(D)$ and can be represented by an exact differential form. □

Our integral computes the period pairing

$$\langle\,,\,\rangle\colon H_1^{\mathrm{sing}}(C^\circ, D) \times H_{\mathrm{dR}}^1(C^\circ, D) \to \mathbb{C},$$

$$(\sigma, \omega) \mapsto \alpha = \langle[\sigma], [\omega]\rangle = \int_\sigma \omega.$$

It vanishes if $[\sigma] = 0$ or $[\omega] = 0$. We are interested in the cases where this condition is not satisfied.

Definition 14.8. We say that the period pairing has *non-trivial vanishing for* (ω, σ) if $\alpha = \langle[\omega], [\sigma]\rangle = 0$ but $[\omega] \neq 0, [\sigma] \neq 0$.

We denote by $\mathcal{N} \subset H_1^{\mathrm{sing}}(C^\circ, D) \times H_{\mathrm{dR}}^1(C^\circ, D)$ the set of pairs $([\sigma], [\omega])$ such that $\langle[\sigma], [\omega]\rangle = 0$. We further denote by $\mathcal{N}^i \subset \mathcal{N}$ for $i = 1, 2, 3$ the subsets for which the differential form is of first, second and third kind, respectively.

Our aim is to determine under which conditions for σ and ω the pairing has non-trivial vanishing.

Note that $D \hookrightarrow D'$ implies an inclusion $H_1^{\mathrm{sing}}(C^\circ, D) \hookrightarrow H_1^{\mathrm{sing}}(C^\circ, D')$ as well as a surjection $H_{\mathrm{dR}}^1(C^\circ, D') \twoheadrightarrow H_{\mathrm{dR}}^1(C^\circ, D)$. Hence vanishing of $[\sigma]$ does not depend on D', but vanishing of $[\omega]$ does.

14.2.2 Translation to Motives

Properties of the periods of $H^1(C^\circ, D)$ depend on the 1-motive

$$M = [L \to J(C^\circ)],$$

where $L = \mathbb{Z}[D]^0$ is the group of divisors of degree 0 supported on D. It has rank r. By Lemma 12.9, more precisely equation (12.1) of its proof, we have

$$V_{\mathrm{sing}}(M) \cong H_1^{\mathrm{sing}}(C^\circ, D; \mathbb{Q}), \quad V_{\mathrm{dR}}^\vee(M) \cong H_{\mathrm{dR}}^1(C^\circ, D),$$

and

$$\alpha = \int_\sigma \omega$$

can be viewed as a period of M.

Lemma 14.9. *The class of σ is given by*

$$[\sigma] = (\partial\sigma, l^\circ(\sigma)) \in T_{\mathrm{sing}}(M) \subset L \times \mathrm{Lie}(J(C^\circ))^{\mathrm{an}}.$$

Both components have image $P(\sigma)$ in $J(C^\circ)$.

Proof This is precisely the identification of classes in $H_1^{\text{sing}}(C^\circ, D; \mathbb{Z})$ with $T_{\text{sing}}(M)$ in Lemma C.5. Classes in relative homology are represented by pairs in $S_0^\infty(D) \oplus S_1^\infty(C^\circ)$, in our case by $(\partial\sigma, \sigma)$. The first summand maps to L, the second maps a path γ to $I(\gamma)$ and hence σ to $I^\circ(\sigma)$. □

The key for determining the spaces $\mathcal{N}$ and $\mathcal{N}^i$ is Theorem 9.7. It shows that non-trivial vanishing of α is caused by a non-trivial exact sequence

$$0 \to M_1 \xrightarrow{\iota} M \xrightarrow{p} M_2 \to 0, \tag{14.1}$$

with $M_i = [L_i \to G_i]$, $[\sigma] = \iota_*\sigma_1$ induced from M_1, $\sigma_1 \in V_{\text{sing}}(M_1)$ and $[\omega] = p^*\omega_2$ induced from M_2, $\omega_2 \in V_{\text{dR}}^\vee(M)$.

To analyse this explicitly, we go through the various types of differential forms. If $J(C)$ is not simple, there are many such sequences and there is a lot of non-trivial vanishing; see Example 14.5(3). The general case seems to be very complicated, so for the rest of this chapter we restrict our discussion to the case where $J(C)$ is simple.

Remark 14.10. By construction, $[\sigma]$ is induced from the submotive

$$M' = [\mathbb{Z}_\sigma \to J(C^\circ)] \subset M,$$

where $\mathbb{Z}_\sigma$ is the sublattice of L generated by the element $\partial\sigma$.

We go now through the different cases, i.e. forms of the first kind, of the second kind and of the third kind, one after the other and analyse the conditions for non-trivial vanishing. This analysis is carried out by going through the different possible shapes of M_1. We finally specialise to the easier case where, in addition, σ is simple.

14.3 Forms of the First Kind

A form of the first kind is a non-zero global differential form $\omega \in H^0(C, \Omega_C^1)$ without pole. This implies that we use $C^\circ = C$. As previously mentioned, we assume for simplicity that $J(C)$ is simple. The class of ω in $H_{\text{dR}}^1(C, D)$ and even in $H_{\text{dR}}^1(C)$ is non-zero because differential forms of the first kind cannot be exact.

In this situation the relevant motive is $M = [\mathbb{Z}[D]^0 \to J(C)]$, and for applying Theorem 9.7 we have to determine the possible submotives M_1. Since $J(C)$ is assumed to be simple these have either the shape $[L_1 \to J(C)]$ or the shape $[L_1 \to 0]$ for some $L_1 \subset \mathbb{Z}[D]^0$ with quotient M_2. We discuss the two cases separately.

14.3.1 The Case $M_1 = [L_1 \to J(C)]$

According to Theorem 9.7, the form $[\omega]$ is a pull-back from M_2, hence its pull-back to $[L_1 \to J(C)]$ is zero, as is its restriction to $[0 \to J(C)]$. This is equivalent to $[\omega] = 0$ in $H^1_{\mathrm{dR}}(C)$. Since it is a differential form of the first kind, this implies $\omega = 0$. This case does not occur.

14.3.2 The Case $M_1 = [L_1 \to 0] \subset M$

Since we are in the category of iso-1-motives, the structure map $L_1 \to J(C)$ is isogenous to the zero map and $\sigma = \iota_*(\sigma_1)$. This means that

$$P(\sigma) = \exp(l(\sigma)) = \sum_i a_i P(\gamma_i) \in J(C)$$

is a torsion point, a necessary but not sufficient condition for non-trivial vanishing. This property has to be translated from a statement about $J(C)$ into a vanishing condition in the Lie algebra. The point is that the exponential map is not injective. This requires a more careful analysis of the situation.

Let $n \geq 1$ be chosen such that the image of L in $J(C)$ under the composition

$$L \to J(C) \xrightarrow{[n]_*} J(C)$$

is torsion free. The kernel of the composition has a complement L_0 with the property that the structure map of the modified motive $M_0 = [L_0 \to J(C)]$ (with structure map the restriction from the structure map of M) is injective. The map $[n]: M \to M$ factors through a morphism of 1-motives $M \to M_0$ with M_1 contained in its kernel. It is multiplication by n on the abelian part. The image of σ is zero in $T_{\mathrm{sing}}(M_0)$ and hence $[n]_* l(\sigma) = 0$ in $\mathrm{Lie}(J(C)^{\mathrm{an}})$. The map $[n]_*$ is multiplication by n on $\mathrm{Lie}(J(C))^{\mathrm{an}}$, hence $[n]_* l(\sigma) = 0$ implies $l(\sigma) = 0$. We conclude that non-trivial vanishing for (σ, ω) implies that $l(\sigma) = 0$, with no condition on ω. In this situation, this means that $\mathcal{N}^1 \subset \ker(l) \times H^0(C, \Omega^1)$.

From $l(\sigma) = 0$ we conclude in particular that $P(\sigma)$ is not only a torsion point but zero. This implies that the points $P(\gamma_i)$ are linearly dependent in $J(C)$.

Conversely, suppose that σ satisfies $l(\sigma) = 0$. We introduce the submotive $M_1 = [\mathbb{Z}_\sigma \to 0]$ with $\mathbb{Z}_\sigma$ the sublattice of $L = \mathbb{Z}[D]^0$ generated by $\partial\sigma$. We map $L_1 = \mathbb{Z}$ to $L = \mathbb{Z}[D]^0$ by mapping 1 to $\sum_{i=1}^n a_i(\gamma_i(1) - \gamma_i(0)) \in \mathbb{Z}[D]^0$ and introduce the motive $M_1 = [L_1 \to 0] \to [L \to J(C)^{\mathrm{an}}]$. It is well defined because $P(\sigma) = 0$ and by construction $[\sigma]$ is induced from a class on M_1.

Any morphism of motives induces a morphism of Hodge structures, which in particular respects the Hodge filtration. By assumption ω is in $H^0(C, \Omega^1) = F^1 H^1_{\mathrm{dR}}(C, D) = F^1 V^\vee_{\mathrm{dR}}(M)$. As a consequence, the pull-back of ω to M_1 is in $F^1 V^\vee_{\mathrm{dR}}(M_1) = 0$ and this gives

$$\int_\sigma \omega = \int_{\iota_* \sigma_1} \omega = \int_{\sigma_1} \iota^* \omega = \int_{\sigma_1} 0 = 0.$$

We conclude that every pair (σ, ω) with $l(\sigma) = 0$ is contained in $\mathcal{N}^1$. In other words, we have $\mathcal{N}^1 \supset \ker(l) \times H^0(C, \Omega^1)$ and we conclude that

$$\mathcal{N}^1 = \ker(l) \times H^0(C, \Omega^1).$$

Summary 14.11. *Let $J(C)$ be simple and ω of the first kind. Then there is non-trivial vanishing for (ω, σ) if and only if $l(\sigma) = 0$. We have $\mathcal{N}^1 = \ker(l) \times H^0(C, \Omega^1)$. This implies in particular that the $I(\gamma_i)$ are $\mathbb{Z}$-linearly dependent in $\mathrm{Lie}(J(C)^{\mathrm{an}})$. A fortiori the points $P(\gamma_i)$ are linearly dependent in $J(C)$.*

Remark 14.12. The condition looks like trivial vanishing because the class of σ vanishes in $H_1^{\mathrm{sing}}(C, \mathbb{Q})$ – but it is not. The class $[\sigma] \in H_1^{\mathrm{sing}}(C, D; \mathbb{Q})$ is not in the image of $H_1^{\mathrm{sing}}(C, \mathbb{Q})$ under the natural restriction. Rather there is a projection $H_1^{\mathrm{sing}}(C, D; \mathbb{Q}) \to H_1^{\mathrm{sing}}(C; \mathbb{Q})$ defined only via the Jacobian. The image of $[\sigma]$ under this projection vanishes.

14.4 Forms of the Second Kind

Assume that ω has a non-empty set of poles but no residues. We have $[\omega] \in H^1_{\mathrm{dR}}(C, D)$ and again $M = [\mathbb{Z}[D]^0 \to J(C)]$. Assume that $J(C)$ is simple and that we have non-trivial vanishing induced from a sub-object $M_1 \subset [L \to J(C)]$. As before, we assume that the period pairing has non-trivial vanishing for (ω, σ). This implies that $\sigma = \iota_* \sigma_1$ and $\omega = p^* \omega_2$.

14.4.1 The Case $M_1 = [L_1 \to J(C)]$

With the same argument as in Section 14.3.1 for ω of the first kind, the assumption on M_1 leads to $\omega = df$ with f not identically zero on D. In the cocycle computation in Section 14.2.1 we can choose all f_i as f and represent the class of $[\omega]$ in $H^1_{\mathrm{dR}}(C, D)$ by

$$\underline{\omega} - \partial f = (0, 0, f|_D).$$

The period integral is in this case

$$\alpha = \langle [\omega], [\sigma] \rangle = \langle f, \partial \sigma \rangle$$
$$= \sum_{i=1}^n a_i (f(\gamma_i(1)) - f(\gamma_i(0))).$$

This is a $\mathbb{Q}$-linear combination of algebraic numbers. It can vanish, and will in examples.

Examples 14.13. 1. Let $E \subset \mathbb{P}^2$ be the elliptic curve given by

$$Y^2 Z = X(X - Z)(X - \lambda Z)$$

for $\lambda \in \overline{\mathbb{Q}} \setminus \{0, 1\}$. Consider $\omega = dX$. This is an exact form. For every $x \in \mathbb{Q}$, there is a point P_x in $E(\overline{\mathbb{Q}})$ with $X(P_x) = x \in \mathbb{Q}$. We choose $x_1, \ldots, x_r \in \mathbb{Q}$. There are coefficients $a_i \in \mathbb{Q}$ such that $\sum a_i x_i = 0$. Let γ_i be a path from P_0 to P_{x_i} and define $\sigma = \sum a_i \gamma_i$. Then $\int_\sigma dX = 0$.

2. Let C be a smooth projective curve and $f \in \overline{\mathbb{Q}}(C)$ not constant. We view f as a non-constant morphism $f \colon C \to \mathbb{P}^1$. For $x \in \mathbb{P}^1(\overline{\mathbb{Q}})$, we choose $y_0, y_1 \in f^{-1}(x)$ and a path γ from y_0 to y_1 in C^{an}. Then $\int_\gamma df = f(y_1) - f(y_0) = x - x = 0$.

Summary 14.14. *If $\omega = df$ and $J(C)$ is simple, then there is non-trivial vanishing if and only if*

$$\langle f, \partial \sigma \rangle = 0. \tag{14.2}$$

If $\sigma = \gamma$ is simple, this is equivalent to $f(\gamma(1)) = f(\gamma(0))$.

14.4.2 The Case $M_1 = [L_1 \to 0]$

The same arguments as for periods of the first kind imply $l(\sigma) = 0$. We show that this is no longer sufficient, but there is also a condition on ω. Let $\mathbb{Z}_\sigma = \mathbb{Z}\partial\sigma \subset \mathbb{Z}[D]^0$ be generated by the divisor $\partial\sigma$. It is contained in L_1 because $l(\sigma) = 0$. Without loss of generality, we take $L_1 = \mathbb{Z}_\sigma$.

The condition on ω to be of the form $p^*\omega_2$ is equivalent to

$$\iota^*([\omega]) = [\omega]|_{[\mathbb{Z}_\sigma \to 0]} = 0. \tag{14.3}$$

Remark 14.15. For any ω of the second kind, the restriction $[\omega]|_{[\mathbb{Z}_\sigma \to 0]}$ is the restriction of an element of $V_{\mathrm{dR}}^\vee([L \to 0])$. Translated back to cohomology of curves, this means that it is in the image of $H_{\mathrm{dR}}^0(D)$ in $H_{\mathrm{dR}}^1(C^\circ, D)$. In Lemma 14.7 the classes in the image are represented by exact differential forms df. In consequence, we find for every ω of the second kind an exact form df such that $\omega - df$ satisfies the vanishing condition. The same is true for the converse: there is always an exact form df such that the vanishing condition is not satisfied.

Summary 14.16. *Let $J(C)$ be simple. Assume that ω is of the second kind, but neither of the first kind nor exact. Then there is non-trivial vanishing if and only if $l(\sigma) = 0$ and equation (14.3) holds.*

14.5 Forms of the Third Kind

Assume that ω is of the third kind but not of the first kind. This means that it has a non-empty set of poles, all of them simple. In the language of Hodge theory, this makes it a differential form with log poles. Here we take $M = [L \to J(C^\circ)]$ as our motive. The form ω defines a class $[\omega]$ in $H^1_{dR}(C^\circ, D)$. It is always non-zero. Hence the period integral vanishes trivially if $[\sigma] = 0$. For simplicity, we assume that $J(C)$ is simple.

14.5.1 The Case $M_1 = [L_1 \to 0]$

As in the other cases, non-trivial vanishing implies $l^\circ(\sigma) = 0$ in $\mathrm{Lie}(J(C^\circ)^{\mathrm{an}})$. We have $\omega \in H^0(C, \Omega^1(\log(S))) = F^1 H^1_{dR}(C^\circ, D)$, so the discussion is identical to the case of differentials of the first kind. The condition is also sufficient. This means that $l^\circ(\sigma) = 0$ implies non-trivial vanishing.

14.5.2 The Case $M_1 = [L_1 \to T_1] \subset [L \to J(C^\circ)]$

Here the quotient is $M_2 = [L_2 \to J(C^\circ)/T_1]$. As σ is in the image of $V_{\mathrm{sing}}(M_1)$, it takes the form $\sigma = \iota_*(\sigma_1)$ for some $\sigma_1 \in V_{\mathrm{sing}}(M_1)$. We deduce that $p_*\sigma = p_*\iota_*\sigma_1 = 0$ and the same is true for its image in $V_{\mathrm{sing}}([L_2 \to J(C)])$, a quotient of M_2. As in the case of differentials of the first kind, this implies that the image $l(\sigma) = I(\nu\sigma)$ of $l^\circ(\sigma) = I(\nu^\circ\sigma)$ in $\mathrm{Lie}(J(C))^{\mathrm{an}}$ vanishes. This means that σ is closed and $l^\circ(\sigma)$ is in $\mathrm{Lie}(T)^{\mathrm{an}}$, where T is the torus part of $J(C^\circ)$ with character group $\mathbb{Z}[S]^0$. In other words, σ defines a homology class of T^{an}. We choose small enough simple loops ε_j in $\mathrm{Lie}(J(C^{\circ,\mathrm{an}}))$ around the singularities $Q_j \in S$ with $l^\circ(\varepsilon_i)$ in $\mathrm{Lie}(T)^{\mathrm{an}}$ and introduce $\varepsilon = \sum n_\sigma(Q_i)\varepsilon_i$, with $n_\sigma(Q_i)$ the winding number of σ around Q_i. The homology classes of ε and σ agree. As a consequence $l^\circ(\sigma) - l^\circ(\varepsilon)$ vanishes in $\mathrm{Lie}(J(C^{\circ,\mathrm{an}}))$. It follows that there is non-trivial vanishing for $\sigma - \varepsilon$ by the case treated in Section 14.5.1. This shows that non-trivial vanishing for σ is equivalent to non-trivial vanishing for ε. The period integral is, up to the factor $2\pi i$, a linear combination of the residues:

$$\int_\sigma \omega = \sum_{i=1}^m 2\pi i \, n_\sigma(Q_i) \mathrm{Res}_{Q_i}(\omega). \qquad (14.4)$$

There are cases where this vanishes, but there are also cases where it does not vanish.

Example 14.17. Choose a small disc Δ_i around each Q_i and let $\gamma_i = \partial\Delta_i$. Then $\int_{\sum \gamma_i} \omega = 0$ by Cauchy's Residue Theorem.

14.5.3 The Remaining Case

Finally we have to consider $M_1 = [L_1 \to G_1]$, where the abelian part of G_1 is $J(C)$. This implies that $G_2 = T_2$ is a torus. Let L_2 be a direct complement of L_1 in L. Then $M_2 = [L_2 \to T_2]$.

By the exact sequence (14.1) in Section 14.2, the pull-back of $[\omega] = p^*\omega_2$ to M_1, and further to $V_{dR}^{\vee}(G_1)$, vanishes. If $G_1 = J(C^\circ)$, then ω would be exact, but ω is of the third kind and cannot be exact. This case does not occur. Therefore $G_1 \subsetneq J(C^\circ)$ and the quotient $J(C^\circ)/G_1 \cong T_2$ is a non-trivial torus. Up to isogeny, the torus of $J(C^\circ)$ decomposes as $T_1 \times T_2$, where T_1 is the torus part of G_1. From the discussion above, we get homomorphisms $J(C^\circ) \to T_2$ and $J(C^\circ) \to G_1$ and as a consequence an isogeny $J(C^\circ) \cong G_1 \times T_2$. The classifying map $X(G_1 \times T_2) \to J(C^\circ)$ is the product of the classifying maps of $X(T_1) \to J(C^\circ)$ and $X(T_2) \to 0$. In particular, $X(T)_{\mathbb{Q}} \cong X(T_1)_{\overline{\mathbb{Q}}} \times X(T_2)_{\mathbb{Q}}$ and the classifying map $X(T) \to J(C)^{\vee}$ vanishes on $X(T_2)_{\overline{\mathbb{Q}}}$.

By Lemma 4.24, the classifying map for the semi-abelian variety $J(C^\circ)$ is

$$X(T) = \mathbb{Z}[S]^0 \to J(C) \cong J(C)^{\vee}.$$

The vanishing of the classifying map on $X(T_2)$ means that there is a linear dependence between the Q_i in $J(C)_{\mathbb{Q}} \cong J(C)_{\mathbb{Q}}^{\vee}$.

The exact sequence (14.1) gives us that the image of $[\sigma] \in H_1^{\mathrm{sing}}(C^\circ, D; \mathbb{Z})$ in $H_1^{\mathrm{sing}}(T_2, D_2; \mathbb{Z})$ is zero, where D_2 is the image of D under the composition of the maps $C^\circ \to J(C^\circ) \to T_2$. Moreover, $[\omega]$ has to be a pull-back from $V_{dR}^{\vee}([L_2 \to T_2])$. In the identification

$$V_{dR}^{\vee}(T) = X(T)_{\overline{\mathbb{Q}}} \cong \overline{\mathbb{Q}}[S]^0,$$

the restriction of ω to $V_{dR}^{\vee}(T)$ is mapped to the divisor $\mathrm{res}(\omega)$ given by

$$\sum \mathrm{res}_{Q_i}(\omega) \, Q_i \in \overline{\mathbb{Q}}[S]^0.$$

This gives the necessary condition, namely that $\mathrm{res}(\omega)$ is in $X(T_2)_{\overline{\mathbb{Q}}}$, or in other words that $\sum \mathrm{res}_{Q_i}(\omega) Q_i = 0$ in $J(C)_{\overline{\mathbb{Q}}}$.

Remark 14.18. The residue condition is necessary but not sufficient. It is preserved by adding a form ω_1 of the first kind, but the restriction to $V_{dR}^{\vee}(G_1)$ changes. The restriction is in $F^1 V_{dR}^{\vee}(G_1) = F^1 V_{dR}^{\vee}(M_1)$ and vanishes in $V_{dR}^{\vee}(T_1)$. Hence it is induced by a form of the first kind. There is always a choice of ω_1 of the first kind such that $\omega - \omega_1$ satisfies the condition for non-trivial vanishing. Conversely, there is also ω_1 such that it does not.

Summary 14.19. *Let $J(C)$ be simple and ω a differential form of the third kind. Then there is non-trivial vanishing for (ω, σ) if and only if one of the following conditions is satisfied:*

1. *$l^\circ(\sigma) = 0$ in $\operatorname{Lie}(J(C^\circ))^{\mathrm{an}}$.*
2. *$l(\sigma)$ in $\operatorname{Lie}(J(C))^{\mathrm{an}}$ vanishes and the linear combination in (14.4) vanishes.*
3. *$\operatorname{res}(\omega)$ sums up to 0 in $J(C)_{\overline{\mathbb{Q}}}$, and there is a quotient-motive $[L' \to T']$ of $[L \to J(C^\circ)]$ with $T' \subset T$ and $X(T') \subset \ker(X(T) \to J(C))$ such that*

   - *$\operatorname{res}(\omega)$ is in $X(T')_{\overline{\mathbb{Q}}}$;*
   - *$[\sigma]$ vanishes in $H_1^{\mathrm{sing}}(T', D'; \mathbb{Z})$, where D' denotes the image of D under the maps $C^\circ \to J(C^\circ) \to T'$;*
   - *$[\omega]$ is a pull-back of some $\omega' \in V_{\mathrm{dR}}^\vee([L' \to T'])$.*

14.6 Arbitrary Differential Forms

So far we have worked out necessary and sufficient conditions for non-trivial vanishing for differential forms of the first, second or third kind. We now turn to the general case and write $\omega = \omega_2 + \omega_3$, with ω_2 of the second and ω_3 of the third kind. Such a decomposition exists, by Lemma 14.2.

Theorem 14.20. *Assume that $J(C)$ is simple and that we have non-trivial vanishing of $\int_\sigma \omega$. Then one of the following conditions is satisfied:*

1. *$l^\circ(\sigma) = 0$, and the vanishing condition of (14.3) holds for ω_2.*
2. *$l(\sigma) = 0$, and for ε as in Section 14.5.2, we have*

$$\int_\varepsilon \omega_3 = \sum_{i=1}^m 2\pi i n_\varepsilon(Q_i) \operatorname{res}_{Q_i}(\omega_2) = 0,$$

   as well as the vanishing condition of (14.3) for ω_2 and $\sigma - \varepsilon$.
3. *$\omega = df$ is exact, and the vanishing condition of (14.2) holds.*
4. *The divisor $\operatorname{res}(\omega)$: $Q_i \mapsto \operatorname{res}_{Q_i}\omega$ on S sums up to 0 in $J(C)_{\overline{\mathbb{Q}}}$, and there is a quotient-motive $[L' \to T']$ of $[L \to J(C^\circ)]$ with $T' \subset T$ and $X(T') \subset \ker(X(T) \to J(C))$ such that*

   - *$\operatorname{res}(\omega)$ is in $X(T')_{\overline{\mathbb{Q}}}$;*
   - *$[\sigma]$ vanishes in $H_1^{\mathrm{sing}}(T', D'; \mathbb{Z})$, where D' denotes the image of D under the maps $C^\circ \to J(C^\circ) \to T'$;*
   - *$[\omega]$ is a pull-back of some $\omega' \in V_{\mathrm{dR}}^\vee([L' \to T'])$.*

   In this case, ω_3 and ω_2 can be chosen such that both period integrals vanish.

The proof is not fully self-contained but partially relies on the structural insights of Part IV.

Proof As discussed in Section 14.2.2, non-trivial vanishing is caused by a short exact sequence

$$0 \to M_1 \overset{\iota}{\to} [\mathbb{Z}[D]^0 \to J(C^\circ)] \overset{p}{\to} M_2 \to 0.$$

We go through the possible cases for M_1.

Case $M_1 = [L_1 \to 0]$. In this case $l^\circ(\sigma) = 0$. By the case of differential forms of the third kind, the condition implies $\int_\sigma \omega_3 = 0$ by Summary 14.19 (1), and the period integral for ω depends only on ω_2. We get the vanishing condition formulated for periods of the second kind.

Case $M_1 = [L_1 \to T_1]$. The discussion starts as for forms of the third kind. We get $l(\sigma) = 0$. We find ε consisting of closed loops around the poles such that $l^\circ(\sigma) = l^\circ(\epsilon)$. The vanishing of period gives an equality for the periods of $\sigma - \epsilon$ (which depends only on ω_2, by the previous case) and the period $l^\circ(\epsilon)$ (which depends only on ω_3 and is a linear combination of residues, as in the case of periods of the third kind). A sufficient condition of vanishing is the vanishing of both summands.

For the converse, consider $\int_\varepsilon \omega_3 = -\int_{\sigma-\varepsilon} \omega_2$. The left-hand side is a $\overline{\mathbb{Q}}$-multiple of $2\pi i$, whereas the right-hand side is an incomplete period of the second kind. By the structural results of Chapter 16, this makes them linearly independent. Both have to vanish. To see this, we view the left-hand side as a Tate period in the terminology of Chapter 16. The right-hand side is an incomplete period of the second kind. By the second part of the linear independence result of Theorem 16.2, both are periods of the first kind with respect to closed paths. By the first linear independence statement of the same theorem, they are in turn linearly independent from Tate periods. This settles the claim.

Case $M_1 = [L_1 \to G_1]$ with abelian part of G_1 equal to $J(C)$. We have $G_1 \subset J(C^\circ)$ with quotient a torus T_2. The pull-back of ω to G_1 is trivial. If $G_1 = J(C^\circ)$, then this makes $T_2 = 0$ and ω exact, a case we have treated in Summary 14.14. This satisfies condition (3) of the theorem.

If $T_2 \neq 0$, then the argument and the conclusion are the same as for forms of the third kind in Section 14.5.3. By Remark 14.18, we can choose a form ω_1 such that $\int_\sigma (\omega_3 - \omega_1)$ vanishes. Hence there is also vanishing for $\omega_2 + \omega_1$. □

14.7 Vanishing of Simple Periods

The discussion simplifies for simple periods. If $\sigma = \gamma$, then $P(\gamma) = 0$ means that the path is closed. The conditions $l(\gamma) = 0$ and $l^\circ(\gamma) = 0$ mean that γ is closed and contractible in $J(C^{\text{an}})$ and even in $J(C^\circ)^{\text{an}}$. In many cases this amounts to trivial vanishing: the homology class of $[\gamma]$ vanishes. We specialise Theorem 14.20 to the case of simple periods.

Corollary 14.21. *Let $J(C)$ be simple. Assume that $\sigma = \gamma$ is simple and choose an arbitrary rational differential form ω. There is non-trivial vanishing of the period $\int_\gamma \omega$ if and only if one of the following condition holds:*

1. *$\omega = df$ is exact and γ links different points in a fibre of $f \colon C^\circ \to \mathbb{A}^1$.*
2. *γ is a closed path, homologous to 0 in C^{an} and winding around the poles of ω in such a way that the linear combination of the residues in (14.4) vanishes.*
3. *Condition (4) of Theorem 14.20 is satisfied.*

Using our algebraicity criterion, this immediately translates into a strong transcendence result for simple periods. For closed γ, we have already seen in Corollary 13.13 that a period is algebraic if and only if it vanishes. Together with the above corollary, we now have a complete picture. We are also ready to treat the case of a non-closed path.

Corollary 14.22. *Let C be a smooth proper curve over $\overline{\mathbb{Q}}$ with simple Jacobian and $\omega \in \Omega^1_{\overline{\mathbb{Q}}(C)}$ a non-zero rational differential form on C. Let γ be a non-closed path on C^{an} avoiding the poles of ω and with endpoints in $C(\overline{\mathbb{Q}})$. Then*

$$\int_\gamma \omega$$

is transcendental, unless one of the following conditions is satisfied:

1. *$\omega = df$ is exact.*
2. *$\omega = df + \phi$ with ϕ of the third kind such that condition (4) of Theorem 14.20 is satisfied for ϕ or, equivalently, for ω.*

Proof Assume that the period is algebraic. By Theorem 13.9 the form ω can be written as $\omega = df + \phi$ with

$$\int_\gamma \phi = 0.$$

We apply what we have learned about the vanishing of simple periods to ϕ. As γ is not closed, its homology class does not vanish. If $[\phi] = 0$, then ϕ is exact. This makes ω exact.

It remains to go through the cases of non-trivial vanishing in Theorem 14.20. Cases (1) and (2) are excluded because γ is not closed. Case (3) gives back exact ϕ.

If condition (4) of Theorem 14.20 is satisfied for ϕ, then $\phi = \phi_2 + \phi_3$ with ϕ_2, ϕ_3 of the second and third kinds, respectively, and

$$\int_\gamma \phi_2 = \int_\gamma \phi_3 = 0.$$

By Summaries 14.14 and 14.16, this can happen only if either ϕ_2 is exact or $l(\gamma) = 0$. The latter is excluded because γ is not closed. The vanishing for ϕ_3 is characterised in Summary 14.19. This is the same condition as in Theorem 14.20. □

PART FOUR

DIMENSIONS OF PERIOD SPACES

15

Dimension Computations: An Estimate

The aim of this chapter is to establish a formula for the dimension of $\mathcal{P}\langle M \rangle$ for any 1-motive M. We have already given a qualitative characterisation in Corollary 9.12. This characterisation is now turned into an explicit quantitative version. Explicit formulas are deduced in terms of the constituents of the motive M. This will first happen under certain simplifying assumptions. The general case will be considered in subsequent chapters.

15.1 Set-up and Terminology

Throughout this chapter let $M = [L \to G]$ be in 1-Mot$_{\overline{\mathbb{Q}}}$ and let $0 \to T \to G \to A \to 0$ be its decomposition into torus and abelian part. In the arguments below we reserve the letter B for simple abelian varieties and define $E(B) = \operatorname{End}(B)_{\mathbb{Q}}$. This is a division algebra of dimension $e(B) = \dim_{\mathbb{Q}} E(B)$. Let

$$A \cong B_1^{n_1} \times \cdots\cdots \times B_m^{n_m}$$

be the isotypical decomposition of A in the category of abelian varieties up to isogeny. As usual, we write $X(T)$ for the character lattice of T. If X is a lattice, we denote by $T(X)$ the corresponding torus. In Section 4.3 we showed that the datum of a semi-abelian variety G is equivalent to the datum of a homomorphism $X(T) \to A^{\vee}(\overline{\mathbb{Q}})_{\mathbb{Q}}$.

To establish a formulation of a dimension formula for the period space becomes rather complicated because there is a non-trivial interplay between the action of $\operatorname{Hom}(A, B)$ on L and the action of $\operatorname{Hom}(A^{\vee}, B^{\vee})$ on the character group of T. The problems disappear in special cases listed below.

Definitions 15.1. We say that the motive M is

1. of *Baker type* if $A = 0$;

2. of *semi-abelian type* if $L = 0$;
3. of *second kind* if $T = 0$;
4. *reduced* if $L \to A(\overline{\mathbb{Q}})_{\mathbb{Q}}$ and $X(T) \to A^{\vee}(\overline{\mathbb{Q}})_{\mathbb{Q}}$ are injective;
5. *saturated* if it is reduced and $\mathrm{End}(M)_{\mathbb{Q}} = \mathrm{End}(A)_{\mathbb{Q}}$.

To state our dimension formulas we need some notation.

Notation 15.2. • Put $\delta(M) = \dim_{\overline{\mathbb{Q}}} \mathcal{P}\langle M \rangle$.
• We define the *L-rank* $\mathrm{rk}_B(L, M)$ *of* M *with respect to* B as the $E(B)$-dimension of the vector space

$$\mathrm{Hom}(A, B) \cdot L := \sum p(L) \subset B(\overline{\mathbb{Q}})_{\mathbb{Q}}$$

for $p \in \mathrm{Hom}(A, B)_{\mathbb{Q}}$, where $p(L)$ denotes the image of L under the composition of the maps $L \to A(\overline{\mathbb{Q}})$ and p.
• The endomorphism algebra $E(B)$ acts on $B^{\vee}$ from the right as $(b^{\vee}, e) \mapsto e^{\vee}(b^{\vee})$, where $e^{\vee}$ signifies the isogeny dual to e. The *T-rank* $\mathrm{rk}_B(T, M)$ *of* M *with respect to* B is the $E(B)$-dimension of the right $E(B)$-vector space

$$\mathrm{Hom}(A^{\vee}, B^{\vee})_{\mathbb{Q}} \cdot X(T) := \sum p(X(T)) \subset B^{\vee}(\overline{\mathbb{Q}})_{\mathbb{Q}}$$

for $p \in \mathrm{Hom}(A^{\vee}, B^{\vee})_{\mathbb{Q}}$, where $p(X(T))$ denotes the image of $X(T)$ under the composition of $X(T) \to A^{\vee}(\overline{\mathbb{Q}})_{\mathbb{Q}}$ with p.
• If $[L \to T]$ is a 1-motive of Baker type, we define the *L-rank* $\mathrm{rk}_{\mathbb{G}_m}(L, M)$ *of* M *in* $\mathbb{G}_m$ as the rank of

$$\mathrm{Hom}(T, \mathbb{G}_m) \cdot L := \sum_{\chi \in X(T)} \chi(L),$$

where $\chi(L)$ denotes the image of L under the composition of $L \to T$ and $\chi \in X(T)$.

The main theorem of the present chapter is an estimate from above of the dimension $\delta(M)$. In many important cases the inequality is even an equality.

Theorem 15.3 (Dimension estimate). *Let M be a 1-motive with constituents as above.*

1. If M is the product of a motive of Baker type M_0 and a saturated motive M_1, then

$$\delta(M) = \delta(T) + \sum_B \frac{4g(B)^2}{e(B)} + \delta(L)$$
$$+ \sum_B (2g(B)\mathrm{rk}_B(T, M) + 2g(B)\mathrm{rk}_B(L, M))$$
$$+ \mathrm{rk}_{\mathbb{G}_m}(L_0, M_0) + \sum_B e(B)g(B)\mathrm{rk}_B(L, M)\mathrm{rk}_B(L, M),$$

where all sums are taken over all simple factors of A, without multiplicities, and $g(B) = \dim B$, $e(B) = \dim_{\mathbb{Q}} \operatorname{End}(B)_{\mathbb{Q}}$.

2. *For every 1-motive M, there is a product $\widetilde{M}$ of a motive of Baker type and a saturated motive $\widetilde{M}$ such that*

$$\mathcal{P}\langle M \rangle \subset \mathcal{P}\langle \widetilde{M} \rangle$$

In particular,

$$\delta(M) \le \delta(\widetilde{M}).$$

The construction is effective (see Lemmas 15.22 and 15.24) in such a way that the abelian parts of M and $\widetilde{M}$ agree and $\operatorname{rk}_B(L, M) = \operatorname{rk}_B(\widetilde{L}, \widetilde{M})$, $\operatorname{rk}_B(T, M) = \operatorname{rk}_B(\widetilde{T}, \widetilde{M})$.

3. *We have $\delta(L) = 1$ if $L \ne 0$ and $\delta(L) = 0$ if $L = 0$, and $\delta(T) = 1$ if $T \ne 0$ and $\delta(T) = 0$ if $T = 0$.*

The proof will be given at the end of the chapter. The following example illustrates the discrepancy between the upper bound and the actual dimension.

Example 15.4. We take for A an elliptic curve E, a non-trivial extension $0 \to \mathbb{G}_m \to G \to E \to 0$ which is non-split up to isogeny and $P \in G(\overline{\mathbb{Q}})$ a point whose image in $E(\overline{\mathbb{Q}})$ is not torsion. We consider the 1-motive

$$M = [\mathbb{Z} \to G]$$

with 1 mapping to P. This is the same motive already considered in Chapter 11.

Case 1. Assume that E does not have complex multiplication, i.e. $\operatorname{End}(E) = \mathbb{Z}$. In this case $B = E$, $g = 1$, $e = e(E) = 1$, $\delta(T) = 1$ and $\delta(L) = 1$. Moreover, $\operatorname{rk}_E(T, M) = 1$ and $\operatorname{rk}_E(L, M) = 1$. The motive is saturated and the theorem predicts

$$\delta(M) = 11.$$

This has already been verified directly in Proposition 11.1.

Case 2. In the case that E has CM by an imaginary quadratic extension $\mathbb{Q}(\tau)$ of $\mathbb{Q}$, i.e. $\operatorname{End}(E)_{\mathbb{Q}} = \mathbb{Q}(\tau)$, there is again a single $B = E$. In this situation, we have $m = 1$, $g = 1$, $e = e(E) = 2$, $\delta(T) = 1$, $\delta(L) = 1$, $\operatorname{rk}_E(L, M) = 1$ and $\operatorname{rk}_E(T, M) = 1$. Going through the construction of $\widetilde{M}$ before Lemma 15.24, we see that we choose the Baker part M_0 as 0 and $\widetilde{M} = M_1$ as saturated. The theorem gives

$$\delta(M) \le 10.$$

By Proposition 11.3, we actually have

$$\delta(M) = 9.$$

Remark 15.5. The example shows that we do not have equality in general. We will refine the formula in Theorem 17.8 and make it completely explicit in Theorem 17.15.

15.1.1 Outline of the Proof

In a first and key step in Section 15.2, we will prove the dimension formula in the saturated case. In order to simplify notation, we will first handle in Section 15.2.1 the case where the abelian part of M is simple, then upgrade to general A.

Section 15.3 is an interlude: we go through the easier cases of Baker motives, motives of semi-abelian type, motives of the second kind and establish the dimension formulas. None of these need to be saturated, but the formulas can be reduced to the saturated case by constructing a saturation with the same periods and ranks.

Finally, in Section 15.4, we wrap up and establish the dimension estimate announced in Theorem 15.3. The case when M is a product of a Baker motive and a saturated motive follows easily with the same type of arguments as in Section 15.2 in the saturated case and in Section 15.3 in the Baker case. Starting with a general motive M, we construct a product of a Baker motive M_0 and a saturated motive M_1 such that the space of periods of M is contained in the space of periods of $M_0 \times M_1$. Here we use the ingredients that we have already identified in the special case.

15.2 The Saturated Case

We start under restrictive assumptions. The saturated case is easier, but enough to deduce all structural properties of the period space.

15.2.1 Saturated and Simple

Assume that M is reduced and saturated (see Definition 15.1) with $A = B$ simple of dimension g. We denote by e the $\mathbb{Q}$-dimension of $E = \mathrm{End}(B)_{\mathbb{Q}}$. Then $V_{\mathrm{sing}}(M)$ is an E-module and we choose an E-basis. To be precise, the inclusions

$$[0 \to T] \subset [0 \to G] \to [L \to G]$$

induce injections

$$V_{\mathrm{sing}}(T) \hookrightarrow V_{\mathrm{sing}}(G) \hookrightarrow V_{\mathrm{sing}}(M),$$

read as inclusions from now on. Let $\sigma = (\sigma_1, \ldots, \sigma_r)$ be an E-basis of $V_{\text{sing}}(T)$. We write $r = |\sigma|$ and use the same conventions for all other pieces. Extend σ by γ to an E-basis of $V_{\text{sing}}(G)$ and by λ to an E-basis of $V_{\text{sing}}(M)$.

In the dual setting, we choose $\overline{\mathbb{Q}}$-bases (sic!) of $V_{\text{dR}}^{\vee}(M)$ along the inclusions

$$V_{\text{dR}}^{\vee}(L) \hookrightarrow V_{\text{dR}}^{\vee}([L \to B]) \hookrightarrow V_{\text{dR}}^{\vee}(M).$$

To be precise, let u be a basis of $V_{\text{dR}}^{\vee}([L \to 0])$. We extend u by ω to a basis of $V_{\text{dR}}^{\vee}([L \to A])$ and the resulting basis by ξ to a basis of $V_{\text{dR}}^{\vee}(M)$. Note that $V_{\text{dR}}^{\vee}(M)$ is also a right E-module, but we are not using this structure at this point.

Our discussion shows that the full period matrix of M with respect to these bases has the shape

$$\begin{pmatrix} \xi(\sigma) & \xi(\gamma) & \xi(\lambda)) \\ \omega(\sigma) & \omega(\gamma) & \omega(\lambda) \\ u(\sigma) & u(\gamma) & u(\lambda) \end{pmatrix} = \begin{pmatrix} \xi(\sigma) & \xi(\gamma) & \xi(\lambda) \\ 0 & \omega(\gamma) & \omega(\lambda) \\ 0 & 0 & u(\lambda) \end{pmatrix}. \tag{15.1}$$

Lemma 15.6. *The entries of the above matrix generate $\mathcal{P}\langle M \rangle$ over $\overline{\mathbb{Q}}$. The elements of $\xi(\sigma)$ are $\overline{\mathbb{Q}}$-multiples of $2\pi i$. The elements of $u(\lambda)$ are $\overline{\mathbb{Q}}$-multiples of 1. These are the only $\overline{\mathbb{Q}}$-linear relations between the entries.*

Proof Let $\alpha_1, \ldots, \alpha_e$ be a $\overline{\mathbb{Q}}$-basis of E. Then the tuples

$$(\alpha_j \sigma, \alpha_j \gamma, \alpha_j \lambda | j = 1, \ldots, e)$$

are a $\mathbb{Q}$-basis of $V_{\text{sing}}(M)$, hence the period space is generated by the complex numbers $\xi_i(\alpha_{j*}\sigma_k), \ldots$ over $\overline{\mathbb{Q}}$. The transformation formula for integrals gives

$$\xi_i(\alpha_{j*}\sigma_k) = (\alpha_j^*\xi_i)(\sigma_k).$$

The class $\alpha_j^*\xi_i$ is a $\overline{\mathbb{Q}}$-linear combination of the basis vectors:

$$\alpha_j^*\xi_i = \sum b_s\xi_s + \sum c_t\omega_t + d_r u_r,$$

which implies, in accordance with the shape of the period matrix, that

$$\xi_i(\alpha_{j*}\sigma_k) = \sum b_s\xi_s(\sigma_k).$$

Similar computations also apply to the rest of the basis and we see that the period space has $\overline{\mathbb{Q}}$-generators as claimed.

Now assume that there is a $\overline{\mathbb{Q}}$-linear relation between the periods. It has the shape

$$\sum a\xi(\sigma) + \sum b\xi(\gamma) + \sum c\xi(\lambda) + \sum d\omega(\gamma) + \sum f\omega(\lambda) + \sum gu(\lambda) = 0. \tag{15.2}$$

Here a is a matrix and the first term stands for the sum $\sum_{i,j} a_{ij}\xi_i(\sigma_j)$. The same convention is used in all other places.

We define $T_\sigma = \mathbb{G}_m^{|\sigma|}$, $M_\sigma = [0 \to T_\sigma]$, $G_\gamma = G^{|\gamma|}$, $M_\gamma = [0 \to G_\gamma]$, $M_\lambda = M^{|\lambda|}$ and consider the product

$$\widetilde{M} = M_\sigma \times M_\gamma \times M_\lambda.$$

It can be written in the form $\widetilde{M} = [\widetilde{L} \to \widetilde{G}]$ with group part

$$0 \to \widetilde{T} \to \widetilde{G} \to \widetilde{A} \to 0$$

for a torus $\widetilde{T}$ and an abelian variety $\widetilde{A}$. We fix elements

$$\widetilde{\gamma} = (\sigma, \gamma, \lambda) \in V_{\text{sing}}(\widetilde{M}),$$
$$\widetilde{\omega} = (\phi, \psi, \vartheta) \in V_{\text{dR}}^\vee(\widetilde{M}),$$

where

$$\phi = \left(\sum a_{ij}\xi_j \,|\, i = 1, \ldots, |\sigma| \right),$$
$$\psi = \left(\sum b_{ij}\xi_j + \sum d_{ij}\omega_j \,|\, i = 1, \ldots, |\gamma| \right),$$
$$\vartheta = \left(\sum c_{ij}\xi_j + \sum f_{ij}\omega_j + \sum g_{ij}u_j \,|\, i = 1, \ldots, |l| \right),$$

using the coefficients of the relation (15.2). Then the $\overline{\mathbb{Q}}$-linear relation (15.2) between the periods implies

$$\widetilde{\omega}(\widetilde{\gamma}) = 0.$$

By the Subgroup Theorem for 1-motives, there is a short exact sequence

$$0 \to M_1 \xrightarrow{\nu} \widetilde{M} \xrightarrow{p} M_2 \to 0,$$

with $M_1 = [L_1 \to G_1]$, $M_2 = [L_2 \to G_2]$ and $\widetilde{\gamma} = \nu_* \gamma_1$, $\widetilde{\omega} = p^* \omega_2$.

We analyse M_2. In a first step we show that $A_2 = 0$. Otherwise assume $A_2 \neq 0$. Then there is a non-zero homomorphism $A_2 \to B$. The composition $\widetilde{A} = B^{|\gamma|} \times B^{|\lambda|} \to A_2 \to B$ is given by a tuple (n, m) of elements of E. We determine the image $\bar{L}_2$ of L_2 in B. We have $\bar{L}_2 = \sum_{i=1}^{|\lambda|} m_i(L)$ because the lattice part of M_γ is 0. The image is contained in L because L is E-stable by saturatedness and $L \to G \to B$ is injective. Here we use the fact that M, being saturated, is reduced. We have $\bar{L}_2 \neq 0$ if there is some $m_i \neq 0$. In this case, m_i is a unit in a division algebra, making the map $m_i \colon L \to L$ bijective. This implies $m_i(L) = L$ and then also $\bar{L}_2 = L$. The image of $\widetilde{\gamma}$ under $\bar{p} \colon \widetilde{M} \to M_2 \to [L \to B]$ is equal to

$$\sum n_j \gamma_j + \sum m_i \lambda_i \in V_{\text{sing}}([L \to B]).$$

It vanishes because $p_* \widetilde{\gamma} = 0$. The image of (γ, λ) is an E-basis of $V_{\text{sing}}([L \to B])$. Therefore $m_i = 0$ for all i and $n_j = 0$ for all j. This contradicts the non-triviality of $\widetilde{A} \to A_2 \to B$. Hence $A_2 = 0$.

Consider the composition of the inclusion of one of the factors $[0 \to G]$ of M_γ into $\widetilde{M}$ with p. It has the shape

$$[0 \to G] \to [L_2 \to T_2].$$

If this map was non-zero, its group component would induce a splitting of G. This is not possible because we have assumed that M is reduced. The map has to vanish. As a consequence, the pull-back of $\widetilde{\omega} = p^*\omega_2$ to G is equal to

$$\widetilde{\omega}|_G = (p^*\omega_2)|_G = 0^*\omega_2 = 0.$$

This means $\widetilde{\omega}|_{M_\gamma} = 0$. In other words, $\psi = 0$. By linear independence of the ξ_j and ω_j this implies $b = 0$, $d = 0$.

We repeat the argument with the inclusion of one of the factors M of M_λ into $\widetilde{M}$. It has the shape

$$[L \to G] \to [L_2 \to T_2].$$

As M is reduced, there is no non-trivial map from G to a torus, so the group component of this map vanishes. The image of L in T_2 has to be 0 and therefore its image in L_2 is in $K_2 = \ker(L_2 \to T_2)$. This is true for all factors M of M_λ, hence the surjection $L^{|\lambda|} \to L_2$ factors via $K_2 \subseteq L_2$. This implies that $L_2 = K_2$ and the structure map $L_2 \to T_2$ of M_2 vanishes. The composition

$$M_\lambda \to \widetilde{M} \xrightarrow{p} M_2 \cong [L_2 \to 0] \times [0 \to T_2]$$

has image in $[L_2 \to 0]$, hence the restriction ϑ of $\widetilde{\omega} = p^*\omega_2$ to $V_{\mathrm{dR}}^\vee(M_\lambda)$ takes values in $V_{\mathrm{dR}}^\vee([L^{|\lambda|} \to 0])$. By linear independence, this implies $c = 0$, $f = 0$.

This shows that our linear relation has been reduced to

$$\sum a\xi(\sigma) + \sum gu(\lambda) = 0.$$

The terms in the first sum are periods of Tate type, hence multiples of $2\pi i$. The terms in the second sum are periods of algebraic type, hence multiples of 1. This reduces the proof to the transcendence of π. This was shown in Corollary 10.1. □

Corollary 15.7. *Let M be saturated, with $A = B$ simple. Then*

$$\delta(M) = \delta(T) + \frac{4g^2}{e} + \delta(L)$$
$$+ 2g\,\mathrm{rk}_B(G, M) + 2g\,\mathrm{rk}_B(L, M) + e\,\mathrm{rk}_B(G, M)\mathrm{rk}_B(L, M).$$

In particular, the formula in Theorem 15.3 holds.

Proof　We read off the numbers from the basis constructed in the proof of the previous lemma. Accordingly we have

$$|\sigma| = \mathrm{rk}_B(G, M), \quad |\gamma| = 2g/e, \quad |\lambda| = \mathrm{rk}_B(L, M),$$
$$|\xi| = e\,\mathrm{rk}_B(G, M), \quad |\omega| = 2g, \quad |u| = e\,\mathrm{rk}_B(L, M). \qquad \square$$

15.2.2　General Saturated Motives

Assume M is saturated, but A is not necessarily simple. We have (up to isogeny)

$$A \cong B_1^{n_1} \times \cdots \times B_m^{n_m},$$

with simple non-isogenous B_i. Hence

$$\mathrm{End}(A)_{\mathbb{Q}} \cong M_{n_1}(E_1) \times \cdots \times M_{n_m}(E_m),$$

with non-isomorphic division algebras E_i.

Lemma 15.8. *There is a natural decomposition*

$$M \cong M_1^{n_1} \times \cdots \times M_m^{n_m},$$

where each M_i is saturated and has abelian part given by B_i.

Proof　Let $p_i \in \mathrm{End}(A)$ be the projector onto $B_i^{n_i}$. This gives a decomposition of the identity $1 = \sum p_i$ into idempotents.

Since M is saturated, we have $\mathrm{End}(A)_{\mathbb{Q}} = \mathrm{End}(M)$, which implies that p_i can also be viewed as a projector on M. We obtain a decomposition

$$M \cong \bigoplus_{i=1}^{m} p_i(M) \tag{15.3}$$

into isotypical components.

We now replace M by one of the factors $p_i(M)$ and drop the index i. The abelian part is now B^n. We write $E = \mathrm{End}(B)_{\mathbb{Q}}$. Let $q_1, \ldots, q_n \in M_n(E)$ be the projections to the components. This gives a decomposition of the identity $1 = q_1 + \cdots + q_n$ into idempotents q_i. As $\mathrm{End}(M) = M_n(E)$, this induces a decomposition

$$M \cong \bigoplus_{i=1}^{n} q_i(M).$$

The permutation matrices in $M_n(E)$ induce isomorphisms between the factors $q_i(M)$, hence we even have

$$M \cong (M')^n,$$

with abelian part of M' given by B. $\qquad \square$

Proposition 15.9. *Consider a saturated* 1-*motive M. Then the formula in Theorem 15.3 holds.*

Proof By Lemma 15.8, we are dealing with the motive

$$M_1^{n_1} \times \cdots \times M_m^{n_m},$$

with M_i as there. We have

$$\mathcal{P}(M_i^{n_i}) = \mathcal{P}(M_i),$$

hence we may without loss of generality assume $n_i = 1$ for $i \geq 1$. We now repeat the proofs of Lemma 15.6 and Corollary 15.7 with an extra index i. □

15.3 Special Cases

The formulas which we have derived so far simplify if one of the constituents of M vanishes. Indeed, what we have shown so far is enough to give not only estimates but complete formulas. It is worth spelling this out explicitly in the different cases.

15.3.1 Motives of Baker Type

The simplest case is the Baker motive $M = [L \to T]$. Here we get back Baker's famous theorem on linear forms in logarithms in its qualitative version.

Proposition 15.10 (Baker's Theorem)**.** *Let* $M = [L \to T]$ *be of Baker type. Then*

$$\delta(M) = \delta(T) + \delta(L) + \mathrm{rk}_{\mathbb{G}_m}(L, M).$$

Proof Let K be the kernel of $L \to T$ and $\bar{L}$ the image (up to torsion). Consider the motivic decomposition $[L \to T] = [K \to 0] \oplus [\bar{L} \to T]$ with $\bar{L} \to T$ injective. The periods of $[\bar{L} \to T]$ agree with the periods of

$$M' = \left[\sum_\chi \chi(L) / \mathrm{Torsion} \to \mathbb{G}_m \right],$$

with χ running through $\mathrm{Hom}(T, \mathbb{G}_m)$, as in the definition of $\mathrm{rk}_{\mathbb{G}_m}(L, M)$. The structure map of M' is injective. The torus part has rank $\mathrm{rk}_{\mathbb{G}_m}(L, T)$. We now choose bases and proceed as in the proofs of Lemma 15.6 and Corollary 15.7, only simpler. □

Remark 15.11. This is precisely Baker's Theorem see [Bak66]; and also [BW07, Theorem 2.3]. It can also be deduced directly from the Analytic Subgroup Theorem in its original form applied to a group of the form $V \times T$ for a vector group of dimension equal to the rank of L. We can even take $\mathbb{G}_a \times T$. This is the line of proof used in [BW07].

15.3.2 The Semi-abelian Case

In this section, let $M = [0 \to G]$ be of semi-abelian type. The dimension computation will be achieved by reduction to the saturated case. For later use, we record the construction of this saturation. As always, G is an extension of an abelian variety A by a torus T. It is determined by

$$X(T) \to A^\vee(\overline{\mathbb{Q}})_\mathbb{Q} = \mathrm{Ext}^1(A, \mathbb{G}_m)_\mathbb{Q};$$

see Corollary 4.11.

Definition 15.12. Let G be a semi-abelian variety, $X = X(T)$ the character group and $E = \mathrm{End}(A)_\mathbb{Q}$. We denote by X_red the image of X in $A^\vee(\overline{\mathbb{Q}})_\mathbb{Q}$ and by G_red the semi-abelian variety defined by $X_\mathrm{red} \to A^\vee(\overline{\mathbb{Q}})_\mathbb{Q}$. Define

$$X_\mathrm{sat} = X_\mathrm{red}E \subset A^\vee(\overline{\mathbb{Q}})_\mathbb{Q}$$

and G_sat as the semi-abelian variety given by $X_\mathrm{sat} \subset \mathbb{A}^\vee(\overline{\mathbb{Q}})_\mathbb{Q}$.

The projection $X \to X_\mathrm{red}$ induces a natural injection

$$G_\mathrm{red} \to G,$$

and the inclusion $X_\mathrm{red} \subset X_\mathrm{red}E$ corresponds to a projection

$$G_\mathrm{sat} \to G_\mathrm{red}.$$

By construction, $\mathrm{End}(G_\mathrm{sat}) = E$. Our next lemma relates the spaces $\mathcal{P}(G)$ and $\mathcal{P}(G_\mathrm{sat})$.

Lemma 15.13. *Let G be semi-abelian. Then*

$$\mathcal{P}\langle G \rangle = \mathcal{P}\langle G_\mathrm{sat} \rangle + \mathcal{P}\langle T \rangle.$$

Proof Let $X' = \ker(X \to A^\vee(\overline{\mathbb{Q}})_\mathbb{Q})$ and T' the torus corresponding to X'. Up to isogeny, this gives a decomposition

$$G \cong T' \times G_\mathrm{red}$$

and deduce that

$$\mathcal{P}\langle G \rangle = \mathcal{P}\langle T' \rangle + \mathcal{P}\langle G_\mathrm{red} \rangle.$$

The first space $\mathcal{P}\langle T' \rangle$ is included in $\mathcal{P}\langle G_{\text{red}} \rangle$ unless $X_{\text{red}} = 0$. In this exceptional case, $G_{\text{red}} = G_{\text{sat}} = A$ and the statement is true. In consequence, it suffices to work in the case $G = G_{\text{red}}$ and we claim that

$$\mathcal{P}\langle G \rangle = \mathcal{P}\langle G_{\text{sat}} \rangle.$$

The surjection $G_{\text{sat}} \to G$ induces an inclusion $\mathcal{P}\langle G \rangle \subset \mathcal{P}\langle G_{\text{sat}} \rangle$ of the period spaces. We now establish the converse inclusion.

Let e be a $\mathbb{Q}$-basis of $E = \text{End}(A)$. (As in Section 15.2.1 we write e for the array $e_1, \dots, e_d$ etc.)

Let σ be a $\mathbb{Q}$-basis of $V_{\text{sing}}(T)$ and extend the basis by γ to a $\mathbb{Q}$-basis of $V_{\text{sing}}(G)$. We write $e_*\sigma$ for the set $e_{i*}\sigma_j$ for $e_i \in e$, $l_j \in l$. The tuple $(e_*\sigma, \gamma)$ is a system of generators of $V_{\text{sing}}(G_{\text{sat}})$.

Let ω be $\overline{\mathbb{Q}}$-basis of $V_{\text{dR}}^{\vee}(A)$ and extend the basis by ξ to a $\overline{\mathbb{Q}}$-basis of $V_{\text{dR}}^{\vee}(G)$. We denote by $e^*\xi$ the array $e_i^*\xi_j$ for $e_i \in e$, $u_j \in u$. Then the array $(\omega, e^*\xi)$ is a system of generators of $V_{\text{dR}}^{\vee}(G_{\text{sat}})$.

All periods of the form $e^*\xi(e_*\sigma)$ are multiples of $2\pi i$, hence contained in $\xi(\sigma)$. The same holds trivially for the periods of the form $\omega(\gamma)$. Consider a period

$$e_i^*\xi_j(\gamma_k) = \xi_j(e_{i*}\gamma_k).$$

The element $e_{i*}\gamma_k \in V_{\text{sing}}(M_{\text{sat}})$ is a $\overline{\mathbb{Q}}$-linear combination of the generators $(e_*\sigma, \gamma)$, hence the period $\xi_j(e_{i*}\gamma_k)$ is a linear combination of periods of the form $\xi(e_*\sigma)$ and $\xi(\gamma)$, both contained in $\mathcal{P}\langle G \rangle$. □

We are now ready for our dimension computation.

Proposition 15.14. *Let G be semi-abelian, an extension of the abelian variety A by a torus T. Then*

$$\delta(M) = \delta(T) + \sum_B \frac{1}{e(B)} 4g(B)^2 + \sum_B 2g(B)\,\text{rk}_B(T, M),$$

where the sum is taken over over the simple factors of A, without multiplicities.

Proof First we consider the exceptional case $G \cong T \times A$, up to isogeny. We have

$$\delta(A) = \sum_B \frac{1}{e(B)} 4g(B)^2$$

as a very special case of Proposition 15.9. In the product case, the claim is the linear independence of the spaces $\mathcal{P}\langle T \rangle$ and $\mathcal{P}\langle A \rangle$. This follows from the dimension formula in the case of a non-trivial extension (or its proof), so we may ignore the exception.

In the non-split case, $\mathcal{P}\langle T \rangle \subset \mathcal{P}\langle G_{sat} \rangle$ and Lemma 15.13 gives

$$\delta(G) = \delta(G_{sat}).$$

In the saturated case, the formula is as special instance of Proposition 15.9. □

Remark 15.15. The statement does not mention 1-motives, and indeed the formula can be deduced directly from the Analytic Subgroup Theorem in its original form.

15.3.3 Motives of the Second Kind

Assume that $M = [L \to A]$ is of second kind. We argue as in the semi-abelian case.

Definition 15.16. Let $M = [L \to A]$ be of second kind, and $E = \mathrm{End}(A)_{\mathbb{Q}}$. We denote by L_{red} the image of L in $A(\overline{\mathbb{Q}})_{\mathbb{Q}}$ and put $M_{red} = [L_{red} \to A]$. Define

$$L_{sat} = EL_{red} \subset A(\overline{\mathbb{Q}})_{\mathbb{Q}}$$

and write M_{sat} for the motive $[L_{sat} \to A]$.

The projection $L \to L_{red}$ induces a natural projection

$$M \to M_{red},$$

and the inclusion $X_{red} \subset X_{red}E$ corresponds to an inclusion

$$M_{sat} \to M_{red}.$$

By construction, $\mathrm{End}(M_{sat})$ is E.

Lemma 15.17. *Let $M = [L \to A]$ be of the second kind. Then*

$$\mathcal{P}\langle M \rangle = \mathcal{P}\langle M_{sat} \rangle + \mathcal{P}\langle L \rangle.$$

Proof Mutis mutandis, the argument is the same as in the proof of the semi-abelian case, for Lemma 15.13. □

Proposition 15.18. *Let $[L \to A]$ be a 1-motive of the second kind. Then*

$$\delta(M) = \delta(L) + \sum_{B} \frac{1}{e(B)} 4g(B)^2 + \sum_{B} 2g(B)\,\mathrm{rk}_B(L, M),$$

where the sum is taken over the simple factors of A, without multiplicities.

Proof Mutis mutandis, the argument is the same as in the semi-abelian case, Proposition 15.14. □

Remark 15.19. In contrast to the semi-abelian case, the argument here uses the language of 1-motives. This can be avoided by considering the vector group $M^{\natural}$ directly. This is not surprising: after all, the Analytic Subgroup Theorem for 1-motives is a consequence of the Analytic Subgroup Theorem in its original form.

There is an alternative argument for the proof: the Cartier dual of $M = [L \to A]$ has the shape $[0 \to G^{\vee}]$, where $G^{\vee}$ is the semi-abelian variety with abelian part $A^{\vee}$ and defined by the classifying homomorphism $L \to (A^{\vee})^{\vee}(\overline{\mathbb{Q}})_{\mathbb{Q}}$. The period spaces of M and $G^{\vee}$ have the same dimension. Moreover, $\text{rk}_B(L, M) = \text{rk}_{B^{\vee}}(T^{\vee}, G^{\vee})$, so the formulas for M and $G^{\vee}$ match.

15.4 Proof of the Dimension Estimate

Theorem 15.3 states a formula for $\delta(M)$ in the case of a product of a motive of Baker type and a saturated motive. We have already handled each factor separately. Looking a little more carefully at the proof will also give the full result.

Proposition 15.20. *Let M be a product of a Baker motive and a saturated 1-motive. Then the dimension formula of Theorem 15.3 holds.*

Proof We have already established the case of saturated motives in Proposition 15.9 and of motives of Baker type in Proposition 15.10. With the same argument as in the proof of Proposition 15.9, we may assume that the saturated motive is of the form

$$M_1 \times \cdots \times M_m,$$

where the M_i have simple abelian part B_i and the B_i are pairwise non-isogenous. The periods remain unchanged. As in the proof of Baker's Theorem, we further may assume that the motive of Baker type is of the form

$$M_0 = [L \to \mathbb{G}_m].$$

The periods and the ranks in the formula remain unchanged by this process.

Going through the argument in the proof of Proposition 15.9, but with $i = 0$, $1, \ldots, m$ instead of $i = 1, \ldots, m$, proves the proposition. □

Since the reduction of the proof of Theorem 15.3 to the case of Proposition 15.20 is lengthy, we offer a short outline of the proof.

Given a 1-motive M we want to find a 1-motive $\widetilde{M}$ with the same abelian part A such that $E = \text{End}(A)$ operates on all of $\widetilde{M}$. The idea is to enlarge both T and

L by adding their E-translates. In the simplest case, the motive is $u\colon \mathbb{Z} \to A$, which is made E-equivariant by replacing $\mathbb{Z}$ by $\mathrm{End}(A)u$. We have already used this device in Definition 15.16. However, this needs first to make M reduced before the procedure can be applied. If this has been achieved, so that we are in the reduced case, we enlarge G to G_{sat} such that E operates on G_{sat}. To this end, we have to choose a lift $L \to G$ to $L \to G_{\mathrm{sat}}$, then enlarge L such that E operates. In some of the steps, the periods remain the same; in other steps, the space of periods is enlarged. In each step the periods can be controlled.

To begin with, the first step is to construct in a canonical way a 1-motive M_{Bk} of Baker type and non-canonically a reduced M_{red} such that

$$\mathcal{P}\langle M\rangle = \mathcal{P}\langle M_{\mathrm{red}}\rangle + \mathcal{P}\langle M_{\mathrm{Bk}}\rangle. \tag{15.4}$$

This is done as follows. The composition $L \to G(\overline{\mathbb{Q}})_{\mathbb{Q}} \to A(\overline{\mathbb{Q}})_{\mathbb{Q}}$ has a kernel L' and an image L''. We have $[L'' \to A] = [L \to A]_{\mathrm{red}}$ with the notation of Definition 15.16. As $L' \to G$ factors via T, this defines a motive of Baker type $[L' \to T]$.

Similarly $X(T) \to A^{\vee}(\overline{\mathbb{Q}})$ has a kernel $X(T'') \subset X(T)$ (for some quotient $T \to T''$) and an image $X(T')$ (for some subtorus $T' \subset T$). As in Remark 4.15, it induces a canonical short exact sequence

$$0 \to G' \to G \to T'' \to 0$$

of semi-abelian groups. The group G' coincides with G_{red} in accordance with the notation of Definition 15.12.

With these data we define a Baker type motive by

$$M_{\mathrm{Bk}} = [L' \to T] \oplus [L \to T''].$$

We now choose splittings $L \cong L' \times L''$ and $X(T) \cong X(T') \times X(T'')$, inducing $G \cong G' \times T''$, all up to isogeny; see Remark 4.15. The composition of $L'' \to G$ with the projection $G \to G'$ defines (uncanonically, depending on the complement) a reduced motive

$$M_{\mathrm{red}} = [L'' \to G'],$$

with the L-rank and the T-rank of M depending only on M_{red}.

Remark 15.21. For later use, we point out the following: if a semi-simple algebra E operates on M, then it will automatically act on M_{Bk}. Moreover, M_{red} can be constructed such that E still operates. We only have to choose the splittings of L and $X(T)$ equivariantly.

Lemma 15.22. *There exists a decomposition of* $\mathcal{P}(M)$ *as*

$$\mathcal{P}\langle M\rangle = \mathcal{P}\langle M_{\mathrm{Bk}}\rangle + \mathcal{P}\langle M_{\mathrm{red}}\rangle.$$

Proof Both M_{Bk} and M_{red} are subquotients of M, hence their periods are contained in $\mathcal{P}(M)$. It remains to check the opposite inclusion.

From Remark 4.15 we know that $G \cong G' \times T''$ with G' reduced as a 1-motive. By composition with the projections, the map $L \to G$ induces $L \to G'$ and $L \to T''$. The induced morphism

$$[L \to G] \to [L \to G'] \times [L \to T'']$$

is injective, hence

$$\mathcal{P}\langle M \rangle \subset \mathcal{P}\langle [L \to G'] \rangle + \mathcal{P}\langle [L \to T''] \rangle.$$

In the second step, by definition of M_{Bk}, we have $L \cong L' \times L''$ with $L'' \to A$ injective and $L' \to G$ factoring via T. The map

$$[L' \to G'] \times [L'' \to G'] \to [L \to G']$$

is surjective, hence

$$\mathcal{P}\langle [L \to G'] \rangle \subset \mathcal{P}\langle [L' \to G'] \rangle + \mathcal{P}\langle [L'' \to G'] \rangle.$$

This is close to the shape we need, but not quite the same. We need to work on the first summand. By assumption, the map $L' \to G$ factors via T, and hence $L' \to G'$ factors via T'. This gives a well-defined morphism

$$[L' \to T'] \times [0 \to G'] \to [L' \to G'],$$

which is surjective. Hence

$$\mathcal{P}\langle [L' \to G'] \rangle \subset \mathcal{P}\langle [L' \to T'] \rangle + \mathcal{P}\langle G' \rangle.$$

Putting these inclusions of period spaces together, we get

$$\mathcal{P}\langle M \rangle \subset \mathcal{P}\langle [L' \to T] \rangle + \mathcal{P}\langle [L'' \to G'] \rangle + \mathcal{P}\langle [L \to T''] \rangle.$$

By definition, the first and the last summand add up to $\mathcal{P}\langle M_{Bk} \rangle$, whereas the middle summand equals $\mathcal{P}\langle M_{red} \rangle$. $\square$

Having finished the reduction step, we now consider the case where M is reduced. We keep $E = \text{End}(A)_{\mathbb{Q}}$. Next we construct a motive $M_{sat} = M_0 \times M_1$ such that $\mathcal{P}\langle M \rangle \subset \mathcal{P}\langle M_{sat} \rangle$, M_0 is of Baker type and M_1 is saturated with $\text{End}(M_1) = E$. To this end, we use T_{sat}, G_{sat}, L_{sat} and $L_{sat} \to A$ such that the E-operation extends to G_{sat} and $[L_{sat} \to A]$; see Definitions 15.12 and 15.16.

It remains to lift $L_{sat} \to A$ to a map $L_{sat} \to G_{sat}$. We choose a lift of $L \to G$ to a morphism $L \to G_{sat}$. The image of $L' := EL \subset G_{sat}(\overline{\mathbb{Q}})$ in $A(\overline{\mathbb{Q}})$ agrees with L_{sat}. By construction, E operates on

$$M' = [L' \to G_{sat}].$$

However, M' is not necessarily reduced (and then saturated) because $L' \to L_{sat}$ is not necessarily injective. We put

$$M_0 = (M')_{Bk}, \quad M_1 = (M')_{red}, \quad M_{sat} = M_0 \times M_1.$$

As pointed out in Remark 15.21, we can choose M'_{red} such that E still operates. This makes M_1 saturated.

Remark 15.23. The construction of M_{sat} depends on the choice of a lift of $L \to G$ to $L \to G_{sat}$. We do not know how to do this in a canonical way.

Lemma 15.24. *Let M be reduced, $M_{sat} = M_0 \times M_1$ as constructed above. Then M_0 is of Baker type, M_1 is saturated and*

$$\mathcal{P}\langle M_{sat} \rangle \supset \mathcal{P}\langle M \rangle.$$

Proof By definition, M_0 is of Baker type and M_1 reduced with $E = \text{End}(A) \subset \text{End}(M_1) \subset \text{End}(A)_{\mathbb{Q}}$, hence M_1 is saturated.

By construction, there is an injection $[L \to G_{sat}] \to M'$ (with M' as in the construction of M_{sat}) and a surjection $[L \to G_{sat}] \to M$. Together this gives the inclusion of period spaces

$$\mathcal{P}\langle M \rangle \subset \mathcal{P}\langle M' \rangle.$$

By Lemma 15.22, we also have

$$\mathcal{P}\langle M' \rangle \subset \mathcal{P}\langle M'_{Bk} \times M'_{red} \rangle = \mathcal{P}\langle M_{sat} \rangle.$$

$\square$

Proof of Theorem 15.3. Part (1) is Proposition 15.20. Part (3) is clear because all periods of $[L \to 0]$ are algebraic and all periods of T are multiples of $2\pi i$. It remains to prove (2). By Lemmas 15.22 and 15.24, we have

$$\mathcal{P}\langle M \rangle \subset \mathcal{P}\langle M_{Bk} \times M_0 \times M_1 \rangle,$$

with M_{Bk} and M_0 of Baker type and M_1 saturated with the same abelian part as M. $\square$

16

Structure of the Period Space

It is not too difficult to determine the structure of the space of periods in the general case by going back to its constituents. The inclusions

$$[0 \to T] \subset [0 \to G] \subset [L \to G] = M$$

induce a filtration

$$V_{\text{sing}}(T) \hookrightarrow V_{\text{sing}}(G) \hookrightarrow V_{\text{sing}}(M)$$

and, from the dual point of view, a cofiltration

$$M = [L \to G] \twoheadrightarrow [L \to A] \twoheadrightarrow [L \to 0],$$

inducing a filtration

$$V_{\text{dR}}^{\vee}(M) \leftarrow V_{\text{dR}}^{\vee}([L \to A]) \leftarrow V_{\text{dR}}^{\vee}([L \to 0]).$$

Together, they introduce a bifiltration

$$
\begin{array}{ccc}
\mathcal{P}\langle T\rangle \;\lhook\joinrel\longrightarrow\; \mathcal{P}\langle G\rangle \;\lhook\joinrel\longrightarrow\; \mathcal{P}\langle M\rangle \\
\big\uparrow \qquad\qquad \big\uparrow \\
\mathcal{P}\langle A\rangle \;\lhook\joinrel\longrightarrow\; \mathcal{P}\langle [L \to A]\rangle \\
\big\uparrow \\
\mathcal{P}\langle [L \to 0]\rangle
\end{array}
$$

on $\mathcal{P}\langle M \rangle$. We introduce some notation (and terminology):

$\mathcal{P}_{\text{Ta}}(M) = \mathcal{P}\langle T \rangle$	(Tate periods),
$\mathcal{P}_2(M) = \mathcal{P}\langle A \rangle$	(2nd kind wrt closed paths),
$\mathcal{P}_{\text{alg}}(M) = \mathcal{P}\langle [L \to 0] \rangle$	(algebraic periods),
$\mathcal{P}_3(M) = \mathcal{P}\langle G \rangle / (\mathcal{P}_{\text{Ta}}(M) + \mathcal{P}_2(M))$	(3rd kind wrt closed paths),
$\mathcal{P}_{\text{inc2}}(M) = \mathcal{P}\langle [L \to A] \rangle / (\mathcal{P}_2(M) + \mathcal{P}_{\text{alg}}(M))$	(2nd kind wrt non-cl. paths),
$\mathcal{P}_{\text{inc3}}(M) = \mathcal{P}\langle M \rangle / (\mathcal{P}_3(M) + \mathcal{P}_{\text{inc2}}(M))$	(3rd kind wrt non-cl. paths).

If M is of Baker type (i.e. $A = 0$), we also use $\mathcal{P}_{\text{Bk}}(M) = \mathcal{P}_{\text{inc3}}(M)$.

From the bifiltration scheme we see that, for example, periods of the third kind with respect to closed paths are only well-defined up to periods of Tate type, and periods of the second kind with respect to closed paths.

Definition 16.1. In each of the cases $? = \text{Ta}, 2, \text{alg}, 3, \text{inc2}, \text{inc3}$, we put

$$\delta_?(M) = \dim_{\overline{\mathbb{Q}}} \mathcal{P}_?(M).$$

The dimensions of the various blocks will be determined one by one. By adding up the $\delta_?(M)$ we then get $\delta(M)$. This works because of the following property.

Theorem 16.2. *The spaces* $\mathcal{P}_{\text{Ta}}(M)$, $\mathcal{P}_{\text{alg}}(M)$ *and* $\mathcal{P}_2(M)$ *have mutually trivial intersection. Moreover,*

$$\mathcal{P}\langle G \rangle \cap \mathcal{P}\langle [L \to A] \rangle = \mathcal{P}\langle A \rangle.$$

Proof These are statements about linear independence. We actually determined bases in the case of a product of a motive of Baker type and a saturated motive in order to determine the dimensions. Also, by Theorem 15.3, we find the periods of M in the space of periods of a certain $\widetilde{M}$ of this special shape. The linear independence claims are simply a by-product.

Alternatively, we can read off the claim from the dimension formulas themselves (rather than their proof). We explain this in detail.

Consider the semi-simple motive

$$M' = [L \to 0] \times [0 \to T] \times [0 \to A].$$

It is the product of a motive of Baker type and a saturated motive. Theorem 15.3 gives

$$\delta(M') = \delta(T) + \delta(L) + \sum_B \frac{1}{e(B)} 4g(B)^2 = \delta_{\text{alg}}(M) + \delta_{\text{Ta}}(M) + \delta_2(M).$$

This means that the period spaces $\mathcal{P}_{\mathrm{Ta}}(M)$, $\mathcal{P}_{\mathrm{alg}}(M)$ and $\mathcal{P}_2(M)$ have mutually trivial intersections.

To prove the second claim, we consider the motive

$$M'' = [L \to A] \times [0 \to G].$$

The first factor is of second kind, the second factor is of semi-abelian type and we have computed the dimension of their period spaces in Propositions 15.14 and 15.18 as:

$$\delta(G) = \delta(T) + \delta(A) + \sum_B 2g(B)\,\mathrm{rk}_B(T, M),$$

$$\delta([L \to A] = \delta(L) + \delta(A) + \sum_B 2g(B)\,\mathrm{rk}_B(L, M);$$

here we have used that $\mathrm{rk}_B(L, M) = \mathrm{rk}_B(L, M'')$ and $\mathrm{rk}_B(T, M) = \mathrm{rk}_B(T, M'')$. On the other hand, combining Lemmas 15.13 and 15.17 on the saturations of motives of semi-abelian type or of the second kind, respectively, we get

$$\mathcal{P}\langle M'' \rangle = \mathcal{P}\langle \widetilde{M}'' \rangle,$$

where

$$\widetilde{M}'' = T \times [L \to 0] \times [0 \to G_{\mathrm{sat}}] \times [L_{\mathrm{sat}} \to A],$$

with G_{sat} and L_{sat} as in Definitions 15.12 and 15.16. This is a product of a motive of Baker type and a saturated motive, hence we find the dimension of its period space by Theorem 15.3 as

$$\delta(M'') = \delta(\widetilde{M}'')$$
$$= \delta(T) + \delta(L) + \delta(A) + \sum_B 2g(B)\,\mathrm{rk}_B(T, M) + \sum_B 2g(B)\,\mathrm{rk}_B(L, M).$$

Moreover, we know that the intersection of $\mathcal{P}\langle G \rangle$ and $\mathcal{P}\langle [L \to A] \rangle$ contains at least $\mathcal{P}\langle A \rangle$. For the numbers to match up, the intersection has to be equal to $\mathcal{P}\langle A \rangle$. □

Remark 16.3. The language of 1-motives has proved useful in the proofs, but it is compulsory for the structural results in the present chapter. The structure of the period space is readily described in terms of the constituents of the 1-motives, but not in terms of finitely generated subgroups of connected commutative algebraic groups and their constituents.

Corollary 16.4. *We always have*

$$\delta(M) = \delta_{\mathrm{Ta}}(M) + \delta_2(M) + \delta_{\mathrm{alg}}(M) + \delta_3(M) + \delta_{\mathrm{inc2}}(M) + \delta_{\mathrm{inc3}}(M).$$

This can be compared with our results for motives of semi-abelian type, of second kind or of Baker type. This is what we know so far:

Proposition 16.5. *1. All Tate periods are $\overline{\mathbb{Q}}$-multiples of $2\pi i$, all algebraic periods are in $\overline{\mathbb{Q}}$. In particular $\delta_{\mathrm{Ta}}(M)$ and $\delta_{\mathrm{alg}}(M)$ take the values 0 or 1, depending on whether T or L are trivial.*
2. We have

$$\delta_2(M) = \sum_B \frac{1}{e(B)} \, 4g(B)^2,$$

where the sum is taken over all simple factors of A, without multiplicities.
3. We have

$$\delta_3(M) = \sum_B 2g(B)\,\mathrm{rk}_B(T, M),$$

$$\delta_{\mathrm{inc2}}(M) = \sum_B 2g(B)\,\mathrm{rk}_B(L, M).$$

4. If M is of the Baker type, then

$$\delta_{\mathrm{inc3}}(M) = \delta_{\mathrm{Bk}}(M) = \mathrm{rk}_{\mathbb{G}_m}(L, M).$$

5. If M is saturated, then

$$\delta_{\mathrm{inc3}}(M) = \sum_B e(B)\,\mathrm{rk}_B(G, M)\,\mathrm{rk}_B(L, M).$$

6. If M is the product of a motive of Baker type and a saturated motive, then the contributions add up.

Remark 16.6. It remains to determine the precise value of $\delta_{\mathrm{inc3}}(M)$ in the general case. This will be carried out in the next chapter. In contrast to the other entries, there does not seem to be an easy and clean answer. The classical cases are misleading in this respect. We are forced to deal with the subtle interplay between the lattice and the torus part, as well as with possible non-trivial endomorphisms of the abelian part.

Translated to algebraic varieties, these are the periods of the third kind with respect to non-closed paths on algebraic curves of genus bigger than 0. Next to nothing was known about them before our monograph. Indeed, as already mentioned before it is here were the point of view of 1-motives really is needed.

17

Incomplete Periods of the Third Kind

In this chapter we develop a precise formula for the dimension $\delta_{\mathrm{inc3}}(M)$ of the space of periods of the third kind with respect to non-closed paths given by

$$\mathcal{P}_{\mathrm{inc3}}(M) = \mathcal{P}\langle M\rangle/(\mathcal{P}\langle G\rangle + \mathcal{P}\langle[L \to A]\rangle),$$

where $M = [L \to G]$ and G is an extension of the abelian variety A by the torus T. It is the most complex part of the picture.

17.1 Relation Spaces

The assignment $M \mapsto \mathcal{P}_{\mathrm{inc3}}(M)$ has only a weak functoriality. If $M' \hookrightarrow M$ is injective or $M \twoheadrightarrow M''$ surjective, the inclusions $\mathcal{P}(M'), \mathcal{P}(M'') \subset \mathcal{P}(M)$ also induce maps

$$\mathcal{P}_{\mathrm{inc3}}(M') \to \mathcal{P}_{\mathrm{inc3}}(M), \quad \mathcal{P}_{\mathrm{inc3}}(M'') \to \mathcal{P}_{\mathrm{inc3}}(M).$$

In terms of the filtration on $\mathcal{P}\langle M\rangle$ we are now interested in the periods of the associated gradeds

$$V_{\mathrm{sing}}(M)/V_{\mathrm{sing}}([0 \to G]) \cong V_{\mathrm{sing}}([L \to 0]),$$
$$V_{\mathrm{dR}}^{\vee}(M)/V_{\mathrm{dR}}^{\vee}([L \to A]) \cong V_{\mathrm{dR}}^{\vee}(T)$$

of the filtrations from Chapter 16 that are of highest degree. We make the identifications

$$V_{\mathrm{dR}}^{\vee}(T) = X(T) \otimes V_{\mathrm{dR}}^{\vee}(\mathbb{G}_m) \cong X(T)_{\overline{\mathbb{Q}}},$$
$$V_{\mathrm{sing}}([L \to 0]) = L \otimes V_{\mathrm{sing}}([\mathbb{Z} \to 0]) \cong L_{\mathbb{Q}}.$$

Given an elementary tensor $l \otimes x \in L \otimes X(T)$, we choose $\xi \in V_{dR}^\vee(M)$ and $\lambda \in V_{sing}(M)$ with image x in $V_{dR}^\vee(T)$ and l in $V_{sing}([L \to 0])$. Then we define a map

$$\Phi \colon L_{\mathbb{Q}} \otimes X(T)_{\mathbb{Q}} \to \mathcal{P}_{inc3}(M),$$

$$l \otimes x \mapsto \xi(\lambda).$$

Lemma 17.1. *The map Φ is well defined and surjective after extension of scalars to $\overline{\mathbb{Q}}$.*

Proof   Two choices of λ differ by an element δ of $V_{sing}(G)$. The period $\xi(\delta)$ depends only on the image of ξ in $V_{dR}^\vee(G)$. Hence it is an element of $\mathcal{P}\langle G \rangle$. The same reasoning works for ξ. Every element of $\mathcal{P}\langle M \rangle$ is a linear combination of elements of the form $\xi(\lambda)$. By definition, we have

$$\xi(\lambda) = \Phi(x \otimes l),$$

where x is the image of ξ in $V_{sing}([L \to 0])$ and l the image of λ in $V_{dR}^\vee(T)$. This makes ϕ surjective.   □

The surjectivity of Φ gives us some initial information for $\delta_{inc3}(M)$. Indeed we have the following result.

Corollary 17.2.

$$\delta_{inc3}(M) \le (\mathrm{rk}L)(\dim T).$$

We also note that the map Φ is compatible with the weak functoriality of $\mathcal{P}_{inc3}$. This can be expressed by the commutative diagram

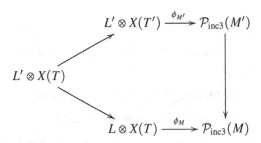

for $M' \hookrightarrow M$ and the same for $M \twoheadrightarrow M''$. In order to determine $\delta_{inc3}(M)$ we need to describe the kernel of Φ.

17.1.1 Structure of Relation Spaces

As a first step we describe the obvious relations and as a second step we verify that they are sufficient.

A morphism of iso-1-motives $\alpha\colon M_1 \to M$ induces morphisms

$$\alpha_*\colon A_1 \to A, \quad \alpha_*\colon L_{1,\mathbb{Q}} \to L_{\mathbb{Q}} \quad \text{and} \quad \alpha_*\colon T_1 \to T.$$

By duality we also get a morphism $\alpha^*\colon X(T)_{\mathbb{Q}} \to X(T_1)_{\mathbb{Q}}$.

Consider an exact sequence

$$M_1 \xrightarrow{\alpha} M \xrightarrow{\beta} M_2.$$

For $l_1 \in L_1$, $x_2 \in X(T_2)$ the period class $\phi_M(\alpha_*(l_1) \otimes \beta^*(x_2))$ agrees with the image of

$$\phi_{M'}(\alpha(l_1) \otimes \beta^* x_2|_{M'}) = \phi_{M'}(\alpha(l_1) \otimes 0) = 0 \in \mathcal{P}_{\mathrm{inc3}}(M'),$$

by the functoriality relation for $M' = \alpha(M_1) \subset M$. This shows that the elements of $\alpha_*(L_1)_{\mathbb{Q}} \otimes \beta^*(X(T_2))_{\mathbb{Q}}$ have image 0 in in $\mathcal{P}_{\mathrm{inc3}}(M)$, which gives us some first relations. They depend on the exact sequence which determines α and β. This suggests to go further and take the sum over all short exact sequences as a first approximation of the relation space.

Definition 17.3. We define the space

$$R_1(M) := \sum_{\alpha,\beta} \alpha_*(L_{1\mathbb{Q}}) \otimes \beta^*(X(T_2)_{\mathbb{Q}}) \subset L_{\mathbb{Q}} \otimes X(T)_{\mathbb{Q}},$$

where the sum is with respect to all exact sequences

$$M_1 \xrightarrow{\alpha} M \xrightarrow{\beta} M_2.$$

Such relations are called *primitive*.

For $n \geq 1$ we extend the definition and introduce the summation maps

$$s_n = \sum p_i \otimes q_i \colon L_{\mathbb{Q}}^n \otimes X(T^n)_{\mathbb{Q}} \to L_{\mathbb{Q}} \otimes X(T)_{\mathbb{Q}},$$

where $p_i\colon V_{\mathrm{sing}}(M^n) \to V_{\mathrm{sing}}(M)$ and $q_i\colon V_{\mathrm{dR}}^{\vee}(M^n) \to V_{\mathrm{dR}}^{\vee}(M)$ are the projections to the ith components, respectively, induced by the projection $\pi_i\colon M^n \to M$ to the ith component and the inclusion $\iota_j\colon M \to M^n$ into the j-component. For $i \neq j$ they satisfy $\pi_i \circ \iota_j = 0$.

Definition 17.4. For $n \geq 1$, we define

$$R_n(M) = s_n(R_1(M^n)).$$

This is justified by the following lemma.

Lemma 17.5. *The space $R_n(M)$ is contained in the kernel of Φ.*

Proof We explained that $R_1(M)$ is contained in the kernel of Φ before we introduced $R_1(M)$. Applying this to the motive M^n we see that

$$\sum_{i,j}(p_i \otimes q_j)(R_1(M^n))$$

is in the kernel of Φ. For $i \neq j$, the orthogonality relation implies that all elements in the image of $p_i \otimes q_j$ are in $\ker(\Phi)$ themselves. Dropping them from the sum, we are still in the kernel. □

Remark 17.6. The space of primitive relations remains unchanged if we restrict our discussion to injective α and surjective β. However, we find the extra flexibility useful. Observe that

$$\beta^*(X(T_2)_\mathbb{Q}) = \ker(\alpha_* : X(T)_\mathbb{Q} \to X(T_1)_\mathbb{Q}),$$
$$\alpha_*(L_{1,\mathbb{Q}}) = \ker(\beta_* : L_\mathbb{Q} \to L_{2,\mathbb{Q}}),$$

where $M_1 = [L_1 \to G_1]$ with torus part T_1, and $M_2 = [L_2 \to G_2]$ with torus part T_2. This gives a less redundant, but also less symmetric, description

$$\begin{aligned} R_1(M) &= \sum_{\alpha:M_1\to M} \mathrm{im}(\alpha_*) \otimes \ker(\alpha^*) \\ &= \sum_{\beta:M\to M_2} \ker(\beta_*) \otimes \mathrm{im}(\beta^*). \end{aligned}$$

There are trivial inclusions $R_n(M) \subset R_{n+1}(M)$. This suggests that we make the following definition.

Definition 17.7. We put

$$R_{\mathrm{inc3}}(M) = \bigcup_{n=1}^\infty R_n(M).$$

The hope is that this makes up all relations. This is the statement of the following theorem.

Theorem 17.8. *The map*

$$(L_\mathbb{Q} \otimes X(T)_\mathbb{Q}/R_{\mathrm{inc3}}(M))_{\overline{\mathbb{Q}}} \to \mathcal{P}_{\mathrm{inc3}}(M)$$

induced by Φ is an isomorphism. In particular, we have

$$\delta_{\mathrm{inc3}}(M) = (\mathrm{rk}L)(\dim T) - \dim(R_{\mathrm{inc3}}(M)).$$

As $L_\mathbb{Q} \otimes X(T)_\mathbb{Q}$ is finite dimensional, the system $R_n(M)$ has to stabilize. Actually, the proof below will show that taking $\mathrm{rk}(L)$ or $\mathrm{rk}(X(T))$ for n suffices. The argument has two steps: first we deal in Section 17.1.2 with the special case $L = \mathbb{Z}$, then in Section 17.1.3 we give the reduction argument.

17.1.2 The Case $L = \mathbb{Z}$

In this section we make the hypothesis that $L = \mathbb{Z}$. In this case $L \otimes X(T)$ is identified with $X(T)$ and with this identification the map Φ simplifies to

$$\Phi \colon X(T)_{\mathbb{Q}} \to \mathcal{P}_{\mathrm{inc3}}(M).$$

Lemma 17.9. *Under this assumption*

$$R_1(M) = \sum_{\alpha} \alpha^* X(T'),$$

where the sum is with respect to all $M' = [L' \to G']$ with torus part T' and all $\alpha \colon M \to M'$ such that $\alpha_*(l) = 0$ for l the image of 1 in $G(\overline{\mathbb{Q}})_{\mathbb{Q}}$.

Proof We use the description

$$R_1(M) = \sum_{\alpha} \ker(\alpha_*) \otimes \mathrm{im}(\alpha^*)$$

with respect to all $\alpha \colon M \to M'$. Clearly we have $\ker(\alpha_*) \neq 0$ if and only if $l \in \ker(\alpha_*)$ and in this case we have identified $\mathbb{Z} \otimes \alpha^* X(T')$ with $\mathrm{im}(\alpha^*)$. $\quad\square$

Proposition 17.10. *Let $M = [\mathbb{Z} \to G]$ be a 1-motive. Then the map*

$$X(T)_{\overline{\mathbb{Q}}}/R_1(M)_{\overline{\mathbb{Q}}} \to \mathcal{P}_{\mathrm{inc3}}(M)$$

is an isomorphism. In particular,

$$R_{\mathrm{inc3}}(M) = R_1(M).$$

Proof We have already discussed surjectivity, so it remains to show

$$\ker(\Phi) \subset R_1(M).$$

We choose bases along the same lines and with the same notation as for the proof of Lemma 15.6, but without taking the $\mathrm{End}(A)$-action into account. This means that $(\sigma, \gamma, \lambda)$ is a $\mathbb{Q}$-basis of $V_{\mathrm{sing}}(M)$ and (u, ω, ξ) is a $\overline{\mathbb{Q}}$-basis of $V_{\mathrm{dR}}^{\vee}(M)$, and moreover, we choose λ such that its image in $V_{\mathrm{sing}}([\mathbb{Z} \to 0])$ is the basis 1. We also choose ξ such that its image $\bar{\xi}$ in $V_{\mathrm{dR}}^{\vee}(T)$ is a $\mathbb{Q}$-basis of $X(T)_{\mathbb{Q}}$.

Suppose we have an element ψ in the kernel of Φ_M. It is of the form $\psi = \sum c_j \xi_j$ such that its period is contained in $\mathcal{P}\langle[L \to A]\rangle + \mathcal{P}\langle G \rangle$ by the definition of $\mathcal{P}_{\mathrm{inc3}}(M)$. Unwinding this we see that there is a $\overline{\mathbb{Q}}$-linear relation of the form

$$\sum a\,\xi(\sigma) + \sum b\,\xi(\gamma) + \sum c\,\xi(\lambda) + \sum d\,\omega(\gamma) + \sum f\,\omega(\lambda) + \sum g\,u(\lambda) = 0.$$

This gives a non-trivial period relation, and the Subgroup Theorem for 1-motives can be applied. We introduce the motives, which can be read off the relation above. They are

$$M_\sigma = [0 \to T_\sigma], \quad T_\sigma = \mathbb{G}_m^{|\sigma|},$$
$$M_\gamma = [0 \to G_\gamma], \quad G_\gamma = G^{|\gamma|},$$
$$M_\lambda = M.$$

We apply the Subgroup Theorem to the motive

$$\widetilde{M} = M_\sigma \times M_\gamma \times M_\lambda = [\widetilde{L} \to \widetilde{G}].$$

In order to write the above relation as a period relation on $\widetilde{M}$, we introduce

$$\widetilde{\gamma} = (\sigma, \gamma, \lambda) \in V_{\mathrm{sing}}(\widetilde{M}),$$
$$\widetilde{\omega} = \left(\sum a_{ij}\xi_j | i = 1, \dots, |\sigma|\right) \times \left(\sum b_{ij}\xi_j + \sum d_{ij}\omega_j | i = 1, \dots, |\gamma|\right)$$
$$\times \left(\sum c_j\xi_j + \sum f_j\omega_j + \sum g_j u_j\right).$$

Then the dependence relation above says that

$$\widetilde{\omega}(\widetilde{\gamma}) = 0.$$

The subgroup theorem for 1-motives gives a short exact sequence

$$0 \to M_1 \xrightarrow{\nu} \widetilde{M} \xrightarrow{p} M_2 \to 0$$

such that $\widetilde{\gamma} = \nu_* \gamma_1$ and $\widetilde{\omega} = p^* \omega_2$. In the next step we unwind what we have obtained so far.

By assumption, the push-forward $p_* \widetilde{\gamma}$ of $\widetilde{\gamma}$ in $V_{\mathrm{sing}}(M_2) = V_{\mathrm{sing}}([L_2 \to G_2])$ vanishes. Further its image in $V_{\mathrm{sing}}([L_2 \to 0]) \cong L_{2,\mathbb{Q}}$ coincides with the image of the generator of $\widetilde{L} \cong \mathbb{Z}$, which vanishes. This implies that $L_2 = 0$ and $M_2 = [0 \to G_2]$.

We now construct the 1-motive M' and $\alpha \colon M \to M'$ as needed in the description of $R_1(M)$ in Lemma 17.9. Let G' be the image of $G_\lambda \subset \widetilde{G}$ in G_2 under $p \colon \widetilde{G} \to G_2$. We write p_λ for the restriction of p to M_λ. The motive $M' = [0 \to G']$ and $\alpha = p_\lambda \colon M = M_\lambda \to M'$ satisfy the conditions for $R_1(M)$. It remains to relate ψ to an element in $\alpha^* X(T')$.

The restriction of $\widetilde{\omega}$ to M_λ depends only on the map $\alpha \colon M_\lambda \to [0 \to G']$ and can be expressed as $\alpha^* \omega'$ for the restriction ω' of ω_2 to $V_{\mathrm{dR}}^\vee(G')$. A further restriction to G_λ gives

$$\widetilde{\omega}|_{G_\lambda} = \sum c_j\xi_j + \sum f_j\omega_j = p_\lambda^* \omega' \in V_{\mathrm{dR}}^\vee(G_\lambda).$$

Restricting to $T_\lambda = T$ gives

$$\widetilde{\omega}|_{T_\lambda} = \sum c_j\xi_j = \alpha^*(\omega'|_{T'}) \in V_{\mathrm{dR}}^\vee(T_\lambda) = X(T)_{\overline{\mathbb{Q}}}.$$

In conclusion, we get

$$\psi = \sum_j c_j \bar{\xi}_j \in \alpha^* X(T')_{\overline{\mathbb{Q}}} \subset R_1(M) \otimes_{\mathbb{Q}} \overline{\mathbb{Q}}.$$

and this proves our claim. □

17.1.3 The Reduction Argument

Given a 1-motive $M = [L \xrightarrow{u} G]$, there is a 1-motive $\widetilde{G}$ with the same periods as M, but a lattice of rank 1: let $l_1, \ldots, l_r$ be a basis of L and put

$$\widetilde{M} = [\mathbb{Z} \to G^r]$$

with structure map $1 \mapsto (u(l_1), \ldots, u(l_r))$. The same choice of basis also induces an identification

$$L \otimes X(T) \cong X(T)^r$$

compatible with the period map Φ.

Lemma 17.11. *In this situation,*

$$R_1(\widetilde{M}) \subset R_1(M^r).$$

Proof Let $\alpha \colon \widetilde{M} \to \widetilde{M}'$ be as in the definition of $R_1(\widetilde{M})$, with M replaced by $\widetilde{M}$; accordingly the image of the basis element $1 \in \mathbb{Z}$ in the lattice part of $\widetilde{M}'$ vanishes. In other words, the map on the group part

$$\alpha \colon G^r \to G'$$

extends to a morphism of motives

$$\alpha \colon [\mathbb{Z} \to G^r] \to [0 \to G'].$$

Let L' be the image of L^r in G' and $M' = [L' \to G']$. Then the induced motive

$$M^r = [L^r \to G^r] \xrightarrow{\alpha'} M'$$

is as in the definition of $R_1(M^r)$. Moreover, the element $(l_1, \ldots, l_r)$ is in the kernel of α' because $\alpha(u(l_1), \ldots, u(l_r)) = 0$. Let $(t_1, \ldots, t_r)$ be an arbitrary element of $\alpha^*(X(T')) \subset R_1(\widetilde{M})$. Then $(l_1, \ldots, l_r) \otimes (t_1, \ldots, t_r)$ fulfils the requirements for being in $R_1(M^r)$. □

Proof of Theorem 17.8. The periods of M, $\widetilde{M}$ and M^r agree. From Proposition 17.10 and the last lemma we know that

$$\ker(\Phi_{\widetilde{M}}) = R_1(\widetilde{M}) \subset R_1(M^r) \subset \ker(\Phi_{M^r}).$$

In consequence, equality holds everywhere. The map Φ_{M^r} factors through the summation map s_r and, by definition, $R_r(M) = s(R_1(M^r))$. This implies

$$\ker(\Phi_M) = R_r(M).$$

As the $R_n(M)$ are nested, this means

$$R_{\text{inc3}}(M) = R_r(M) = \ker(\Phi_M). \qquad \square$$

Corollary 17.12. *We have*

$$R_{\text{inc3}}(M) = R_n(M),$$

where n is the minimum of $\text{rk} L$ *and* $\dim T$.

Proof The proof of the theorem gave equality for $n = r$. The dual arguments allows reduction to the case where the torus is of dimension 1 and hence $n = \dim T$ is also enough. $\qquad \square$

17.2 Alternative Description of $\delta_{\text{inc3}}(M)$

In the last section, we used a trick to reduce the computation of $\delta_{\text{inc3}}(M)$ to the case of motives with lattice part of rank 1. The trick has a canonical interpretation that will lead to an alternative description of $R_{\text{inc3}}(M)$.

Lemma 17.13. *Let $\mathcal{A}$ be an additive category and $\mathbb{Z}$–Proj the category of finitely generated free abelian groups of finite rank. There is an additive bifunctor, the external tensor product*

$$\otimes \colon \mathbb{Z}\text{–Proj} \times \mathcal{A} \to \mathcal{A},$$

uniquely determined by $(\mathbb{Z}, X) \mapsto X$.

Proof Let Λ be a free $\mathbb{Z}$-module of rank r and choose a basis $\lambda_1, \dots, \lambda_r$. For $X \in \mathcal{A}$ we put

$$\Lambda \otimes X := X^r.$$

Let $f \colon \Lambda \to \Lambda'$ be a $\mathbb{Z}$-linear map. In our chosen bases it is given by a matrix $(a_{ij})_{i,j}$ with entries in $\mathbb{Z}$. We define

$$f_* \colon \Lambda \otimes X \to \Lambda' \otimes X$$

as the map $X^r \to X^{r'}$ defined by the matrix. $\qquad \square$

The construction is applied to the categories of abelian or semi-abelian varieties. Given a 1-motive $M = [L \xrightarrow{u} G]$ we can consider $\Lambda = L^\vee$ and the semi-abelian variety $L^\vee \otimes G \cong G^r$. The structure map $u: L \to G$ of M induces a canonical homomorphism $\mathbb{Z} \to L^\vee \otimes G$ as follows: if $l_1, \ldots, l_r$ is a basis of L, then

$$1 \mapsto c := \sum_i l_i^\vee \otimes u(l_i)$$

is the image of 1. We refer to c as the *tautological element*. This map is independent of the choice of basis and has the properties of an adjoint of u.

Remark 17.14. The 1-motive $[\mathbb{Z} \to L^\vee \otimes G]$ agrees with the 1-motive $\widetilde{M}$ considered in Section 17.1.3.

The abelian part of $[\mathbb{Z} \to L^\vee \otimes G]$ is $L^\vee \otimes A$, and the torus part is $L^\vee \otimes T$ with character group $X(L^\vee \otimes T) = L \otimes X(T)$.

As pointed out earlier, the periods of M agree with the periods of our adjoint. Moreover, the map Φ for $[\mathbb{Z} \to L^\vee \otimes G]$ given by

$$\Phi: \mathbb{Z} \otimes X(L^\vee \otimes T) = L \otimes X(T) \to \mathcal{P}_{inc3}(M)$$

agrees with the map Φ_M. We have shown in Proposition 17.10 that

$$\ker(\Phi) = R_1([\mathbb{Z} \to L^\vee \otimes G]).$$

Our objective is to give an explicit description of this space.

The algebra $E = \mathrm{End}(L^\vee \otimes A)_\mathbb{Q}$ operates (up to isogeny) from the right on $(L^\vee \otimes A)^\vee \cong L \otimes A^\vee$. On the other hand, the semi-abelian variety $L^\vee \otimes G$ is characterised by a homomorphism $[L^\vee \otimes G]: X(L^\vee \otimes T) \to (L^\vee \otimes A)^\vee(\overline{\mathbb{Q}})_{\overline{\mathbb{Q}}}$. A choice of elements $\alpha \in E$, $y \in L \otimes X(T)$, $x \in L \otimes A^\vee(\overline{\mathbb{Q}})_\mathbb{Q}$ determines extensions G_x, $G_{\alpha^\vee(x)} \cong \alpha^* G_x$ and $G_{[L^\vee \otimes G](y)}$ in $\mathrm{Ext}^1(L^\vee \otimes A, \mathbb{G}_m)$.

We require that the following compatibility conditions are satisfied.

(A) $\alpha^\vee(x) = [L^\vee \otimes G](y)$; in other words, the diagram

$$
\begin{array}{ccc}
X(L^\vee \otimes T) & \xleftarrow{\ y \leftarrow 1\ } & \mathbb{Z} = X(\mathbb{G}_m) \\
{\scriptstyle [L^\vee \otimes G]}\big\downarrow & & \big\downarrow{\scriptstyle [G_x]} \\
L \otimes A^\vee(\overline{\mathbb{Q}}) & \xleftarrow[\ \alpha^\vee\ ]{} & L \otimes A^\vee(\overline{\mathbb{Q}})
\end{array}
$$

is commutative, which means that α extends to a morphism

$$\alpha_y: L^\vee \otimes G \to G_x.$$

(B) $\alpha_y(c) = 0$ in G_x and again as a consequence α_y defines a morphism

$$[\mathbb{Z} \to L^\vee \otimes G] \to [0 \to G_x].$$

Theorem 17.15. *Let $M = [L \xrightarrow{u} G]$ be a 1-motive. Then*

$$R_{\mathrm{inc3}}(M) = \big\langle y \in L_\mathbb{Q} \otimes X(T)_\mathbb{Q} \mid$$
$$\exists \alpha \in \mathrm{End}(L^\vee \otimes A)_\mathbb{Q}, \exists x \in (L \otimes A^\vee)(\overline{\mathbb{Q}})_{\overline{\mathbb{Q}}}, (A), (B)\big\rangle.$$

Proof We have to check that the set on the right coincides with $R_1([\mathbb{Z} \to L^\vee \otimes G])$. Given $\alpha \in E$ and x with (A), the morphism α extends to $\alpha_y : L^\vee \otimes G \to G_x$. The element $y \in X(L^\vee \otimes T)$ is in the image of $\widetilde{\alpha}^*$ by (A), hence a primitive relation.

Conversely, take $y \in R_1([\mathbb{Z} \to L^\vee \otimes G])$. By Lemma 17.9 there are $\alpha : [\mathbb{Z} \to L^\vee \otimes G] \to M'$ and $x \in X(T')$ such that $y = \alpha^*(x)$ (this equality is property (A)) and the tautological element c is in $\ker(\alpha_*)$. A fortiori, the image of c vanishes in $X(G_x)$ (the vanishing is property (B)).

The abelian part A' of M' is a quotient of $L^\vee \otimes A$. By semi-simplicity, we can choose a direct complement A''. By abuse of notation, we also denote the projector $L^\vee \otimes A \to A' \to L^\vee \otimes A$ obtained in this way by α. We may now replace M' by $M' \times [0 \to A'']$. The new data define an element of the right-hand side. $\square$

Example 17.16. We go back to Example 15.4 and take for A an elliptic curve E, $0 \to \mathbb{G}_m \to G \to E \to 0$ a non-trivial extension (non-split, even up to isogeny) and $P \in G(\overline{\mathbb{Q}})$ a point whose image in $E(\overline{\mathbb{Q}})$ is not torsion. We consider

$$M = [\mathbb{Z} \xrightarrow{u} G]$$

with $u(1) = P$. The surjection

$$\Phi \colon \mathbb{Z} \otimes X(\mathbb{G}_m) = \mathbb{Z} \to \mathcal{P}_{\mathrm{inc3}}(M)$$

gives an upper bound of $\delta_{\mathrm{inc3}}(M) \le 1$. It remains for us to compute the relations. Assume that $0 \ne y \in R_{\mathrm{inc3}}(M)$. By Theorem 17.15, this means there are $\alpha \in \mathrm{End}_\mathbb{Q}(E)_\mathbb{Q}$ and $x \in E^\vee(\overline{\mathbb{Q}})_{\overline{\mathbb{Q}}}$ such that $\alpha^\vee(x) = [G](y)$ and $\alpha_y(c) = 0$ in G_x. The first condition implies that α and x are different from zero. This makes α invertible and we get an isomorphism $G \to G_x$. The tautological class $c \in G(\overline{\mathbb{Q}})_\mathbb{Q}$ is equal to P in our case. It was assumed to be non-torsion in $G(\overline{\mathbb{Q}})$, and the same remains true after applying the isomorphism α. We get a contradiction to (B), and this shows that $R_{\mathrm{inc3}}(M) = 0$ and

$$\delta_{\mathrm{inc3}}(M) = 1.$$

This fits with our explicit computation in Chapter 11, both without CM and with CM.

Remark 17.17. We had suspected that there might be a better description of $R_{\mathrm{inc3}}(M)$ in the language of biextensions. We now tend to think that this is not the case. The period pairing is not related to the pairing between a 1-motive and its Cartier dual.

18

Elliptic Curves

In some sense, Baker's theory of linear forms in logarithms can be seen as an intermezzo, although one of the most influential, in establishing a modern theory of periods. We will now describe a second very important aspect of the theory, namely elliptic periods, which has been developed in the last hundred years by many authors, starting with Siegel and Schneider. We will describe it first in a classical way as has been understood by these authors and then give the translation into our modern language of 1-motives. For more details about the history, see the introduction, in particular Section 1.3.

18.1 Classical Theory of Periods

We review the classical theory of elliptic curves from an algebraic and analytic point of view. This will be used for an application of our abstract results about 1-motives to explicit transcendence results in the elliptic case. For basics on elliptic curves and functions, we refer the reader to [Ahl53, Chapter 7] or [Cha85, Chapters III, IV].

Let E be an elliptic curve given in the projective plane $\mathbb{P}^2$ by an equation of the form

$$y^2 w = 4x^3 - g_2 xw^2 - g_3 w^3, \tag{18.1}$$

with complex parameters g_2 and g_3 such that the discriminant $\Delta = g_2^3 - 27g_3^2 \neq 0$.

Addition can be defined on E, and one obtains a projective commutative algebraic curve with unit element $e_\infty = [0: 0: 1]$, one of the four zeroes e_∞, e_1, e_2, e_3 of the right-hand side of (18.1). These are the four Weierstraß points, the 2-torsion points on the curve.

The associated complex manifold E^{an} becomes a complex Lie group with Lie algebra $\mathrm{Lie}(E)$ and exponential map

$$\exp_E \colon \operatorname{Lie}(E) \to E^{\mathrm{an}}$$

with kernel a lattice Λ, which leads to an exact sequence

$$0 \to \Lambda \to \mathbb{C} \xrightarrow{\exp_E} E^{\mathrm{an}} \to 0$$

and which shows that $E^{\mathrm{an}} \simeq \mathbb{C}/\Lambda$.

In terms of complex analysis, this can be described as follows. For a pair of complex numbers ω_1, ω_2 with $\tau = \omega_2/\omega_1$ in the upper half plane $\Im\tau > 0$, we write $\Lambda = \mathbb{Z}\omega_1 + \mathbb{Z}\omega_2$ and consider the *Weierstraß elliptic function*

$$\wp(z; \Lambda) = \frac{1}{z^2} + \sum_{0 \neq \omega \in \Lambda} \left[\frac{1}{(z - \omega)^2} - \frac{1}{\omega^2} \right].$$

The Weierstraß elliptic function is meromorphic, with poles of order 2 on the lattice Λ, and periodic with period lattice Λ. For $z \in \mathbb{C}$ the triple $(w, x, y) = (1, \wp(z; \Lambda), \wp'(z; \Lambda))$ satisfies equation (18.1) with

$$g_2 = g_2(\Lambda) = 60 \sum_{\omega \neq 0} \frac{1}{\omega^2} \quad \text{and} \quad g_3 = g_3(\Lambda) = 140 \sum_{\omega \neq 0} \frac{1}{\omega^3}.$$

For $z \notin \Lambda$ in $\mathbb{C}$, the exponential map $\exp_E$ can be written as

$$\exp_E(z) = [1 \colon \wp(z) \colon \wp'(z)]$$

in terms of the Weierstraß $\wp$-function. It parametrises the plane algebraic curve

$$y^2 = 4x^3 - g_2 x - g_3.$$

In terms of the uniformisation by the exponential map, the Weierstraß points are $e_\infty = \exp_E(0)$, $e_1 = \exp_E(\omega_1/2)$, $e_2 = \exp_E(\omega_2/2)$ and $e_3 = \exp_E((\omega_1 + \omega_2)/2)$ for a basis ω_1, ω_2 of Λ.

There are two more classical Weierstraß functions which are derived from the *Weierstraß $\wp$-function*. The first is the *Weierstraß ζ-function*

$$\zeta(z; \Lambda) = \frac{1}{z} + \sum_{0 \neq \omega \in \Lambda} \left[\frac{1}{(z - \omega)} + \frac{1}{\omega} + \frac{z}{\omega^2} \right],$$

and the second the *Weierstraß σ-function*

$$\sigma(z; \Lambda) = z \prod_{0 \neq \omega \in \Lambda} \left(1 - \frac{z}{\omega} \right) e^{\frac{z}{\omega} + \frac{1}{2}\left(\frac{z}{\omega}\right)^2}.$$

The latter can be seen as a variant of the Jacobi theta-function, which makes up Jacobi's theory of elliptic functions. The three functions are related by the differential equations

$$\frac{d}{dz}\log(\sigma(z;\Lambda)) = \zeta(z;\Lambda) \quad\text{and}\quad \frac{d}{dz}\zeta(z,\Lambda) = -\wp(z;\Lambda).$$

For $u \in \mathbb{C}$ fixed, we put

$$F(z;u) = \frac{\sigma(z-u)}{\sigma(z)\sigma(u)}\, e^{\zeta(u)z}$$

and compute that

$$\frac{d\log F(z;u)}{dz} = \zeta(z-u) - \zeta(z) + \zeta(u).$$

The classical differential forms of the first, second and third kind are

$$\omega = \frac{dx}{y}, \quad \eta = \frac{x\,dx}{y} \quad\text{and}\quad \xi_P = \frac{y+y(P)}{x-x(P)}\frac{dx}{y}, \tag{18.2}$$

where $P = \exp_E(u)$ is fixed.

The functions x and y have polar divisors $2(e_\infty)$ and $3(e_\infty)$ respectively, so that the differential form ω is regular and η has a pole of order 2 at e_∞. The function $x - x(P)$ has a zero at P and at $-P$ because it is even, and $y + y(P)$ is zero at $(-P)$ and at two other points P_1, P_2. This shows that the divisor of ξ_P is

$$(\xi_P) = ((P_1) + (P_2)) - ((e_\infty) + (P)).$$

We conclude that the polar divisor of ξ_P is $(e_\infty) + (P)$ with residue -2 at (e_∞) and 2 at (P).

Using [Cha85, Chapter IV, §3, equation (3.6)], one verifies that

$$d\log F(z;u) = (\zeta(z-u) - \zeta(z) + \zeta(u))\,dz$$
$$= \frac{1}{2}\frac{\wp'(z) + \wp'(u)}{\wp(z) - \wp(u)}\,dz = \exp_E^* \xi_P.$$

This gives

$$\exp_E^* \omega = dz, \quad \exp_E^* \eta = -d\zeta(z) \quad\text{and}\quad \exp_E^* \xi_P = d\log F(z,u).$$

The Weierstraß ζ-function is quasi-periodic with quasi-periods $\eta_i = 2\zeta(\omega_i/2)$ as is readily seen. Note that this agrees with the normalisation of [Ahl53, Section 3.2] and [Fri11] but differs from [Cha85, Chapter IV, §1, Theorems 1 and 3]. The Weierstraß functions transform as

$$\wp(z + \omega_i) = \wp(z),$$

$$\zeta(z + \omega_i) = \zeta(z) + \eta_i,$$

$$\sigma(z + \omega_i) = -\sigma(z)\, e^{\eta_i\left(z + \frac{\omega_i}{2}\right)},$$

$$F(z + \omega_i, u) = F(z, u)\, e^{-\eta_i u + \zeta(u)\omega_i} = F(z, u)\, e^{\lambda(u, \omega_i)},$$

where $\lambda(u, \omega_i) = \zeta(u)\omega_i - \eta_i u$ for $u \notin \Lambda$. Again one reads off that the function $\wp$ is periodic, as already mentioned, and that the functions ζ, σ and F are quasi-periodic with periods η_i, $e^{\eta_i\left(z + \frac{\omega_i}{2}\right)}$ and $e^{\lambda(u, \omega_i)}$. The function λ can be extended additively to a function on $\mathbb{C} \times \Lambda$. Note that we have $\lambda(\omega_1/2, \omega_2) = \pi i$ by the Legendre relation.

18.2 Elliptic Periods

For $Q = \exp_E(v) \in E^{\mathrm{an}}$ and γ a path from e_∞, to Q the integral

$$\omega(\gamma) = \int_{e_\infty}^{Q} \omega = \int_0^v \exp_E^* \omega = \int_0^v dz = v$$

defines a multi-valued map from E^{an} to $\mathbb{C}$. For different paths from e_∞ to Q the integrals differ by a period $\omega \in \Lambda$. We get back the generators of Λ as the periods

$$\omega_1 = \omega(\varepsilon_1) = \int_{\varepsilon_1} \omega \quad \text{and} \quad \omega_2 = \omega(\varepsilon_2) = \int_{\varepsilon_2} \omega$$

taken along the basis $\varepsilon_1, \varepsilon_2$ of $H_1^{\mathrm{sing}}(E^{\mathrm{an}}, \mathbb{Z})$ defined as the image of the straight paths $[0, \omega_i]$ in $\mathbb{C}$. The integral $\omega(\gamma)$ is called the *incomplete period* of the first kind and becomes a period (i.e. an element of Λ) if $\gamma(1) = e_\infty$.

In the case of periods of the second kind, the path γ must not contain the pole of η, i.e. the Weierstraß point $0 = e_\infty$. For the closed paths ε_1 and ε_2 we get back the quasi-periods $\eta_i = \eta(\varepsilon_i)$ from above.

For the differentials of the third kind ξ_P as above, with polar divisor $(e_\infty) + (P)$, we have to consider in addition a closed path ε_0 going once counterclockwise around 0 with no other singularities inside, and then $\lambda(u, \omega_i)$ and $2\pi i$ become complete periods of the third kind.

In the case of incomplete periods of the second kind, we take a path γ with $\gamma(0) = \exp_E(v)$ and $\gamma(1) = \exp_E(v + w)$ which does not pass through e_∞ and obtain

$$\eta(\gamma) = \int_\gamma \eta = \int_v^{v+w} \exp_E^*(\omega) = \int_v^{v+w} \wp(z)dz = -\zeta(w+v) + \zeta(v)$$

$$= -\zeta(w) - \frac{1}{2}\frac{\wp'(w) - \wp'(v)}{\wp(w) - \wp(v)}$$

$$= -\zeta(w) + \alpha(\gamma).$$

by [Fri11, (2) on p. 202]. When E is defined over $\overline{\mathbb{Q}}$, $\alpha(\gamma)$ is algebraic.

Let $\gamma: [0,1] \to E^{\mathrm{an}}$ be a path of the form $\exp_E \circ \delta$ with $\delta: [0,1] \to \mathrm{Lie}\, E^{\mathrm{an}}$ from v to w continuously differentiable and not containing any of the two poles of the differential of the third kind ξ_P. Then its period along γ is

$$\xi_P(\gamma) = \int_\gamma \xi_P = \int_{\exp \circ \delta} \xi_P = \int_\delta \exp^* \xi_P = \int_\delta \frac{F'(z,u)}{F(z,u)}dz$$

$$= \int_\delta \frac{1}{F(z,u)}dF(z,u) = \int_{F \circ \delta} \frac{dt}{t}$$

$$= \log_{F \circ \delta}(F \circ \delta(1)) - \log(F \circ \delta(0)),$$

where $\log_{F \circ \delta}$ is the branch of the logarithm defined by analytically continuing the function $\log$ from the starting point $F \circ \delta(0) = F(v,u)$ along $F \circ \delta$. Any two branches of the logarithm differ by an integral multiple of $2\pi i$. Up to such multiples, $\log$ satisfies the usual functional equation, hence

$$\xi_P(\gamma) = \log \frac{F(w,u)}{F(v,u)} + 2\pi i v$$

for some $v \in \mathbb{Z}$.

In both cases, we see that if γ is closed with period ω, we get complete periods $\eta(\gamma)$ and $\xi_P(\gamma) = \lambda(u, \omega) + 2\pi i v$.

Summary 18.1. *Let* $\gamma: [0,1] \to E^{\mathrm{an}}$ *be a path. We write it as* $\exp_E \circ \delta$, *where* δ *is a path in* $\mathbb{C} \cong \mathrm{Lie}(E)^{\mathrm{an}}$ *with* $\delta(0) = v$, $\delta(1) = v + w$. *For* ω, η, ξ_P *as in* (18.2) *and* $P = \exp_E(u)$, *we get*

$$\omega(\gamma) = w,$$

$$\eta(\gamma) = -\zeta(w) - \frac{1}{2}\frac{\wp'(w) - \wp'(v)}{\wp(w) - \wp(v)},$$

$$\xi_P(\gamma) = \log \frac{F(v+w, u)}{F(v,u)} + 2\pi i v(\gamma).$$

Let $\varepsilon_1, \varepsilon_2$ *be the generators of* $H_1(E^{\mathrm{an}}, \mathbb{Z})$. *Then* $\omega(\varepsilon_i) = \omega_i$ *are the generators of the period lattice of* E^{an}, $\eta(\varepsilon_i) = \eta_i$ *are the quasi-periods of* E^{an} *and* $\xi_P(\varepsilon_i) = \lambda(u, \omega_i) + 2\pi i v$, *for some* $v(\gamma) \in \mathbb{Z}$, *are the periods of the third kind.*

The period computation for these special differential forms extends to all differential forms.

Lemma 18.2. *Every meromorphic differential form ϑ on E can be written as*

$$\vartheta = a\omega + b\eta + \sum_i c_i \xi_{P_i} + df,$$

with complex coefficients $P_i \in E$ and elliptic f.

Proof Let $P_1, \ldots, P_k$ be the points different from e_∞ where ϑ has a non-vanishing residue c_i. Then $\vartheta' = \vartheta - \sum_i c_i \xi_{P_i}$ has vanishing residues in all points different from e_∞. As the sum of all residues vanishes, it is even of the second kind. As spelt out in Section 14.2.1, it defines a class in $H^1_{dR}(C)$. On the other hand, the classes of ω, ϑ are a basis of the same cohomology group. This gives rise to the identity

$$[\vartheta'] = a[\omega] + b[\eta]$$

for suitable $a, b \in \mathbb{C}$ and makes $\vartheta' - a\omega - b\eta$ exact. □

Clearly if E and ϑ are defined over $\overline{\mathbb{Q}}$, then everything can also be chosen over $\overline{\mathbb{Q}}$. By the lemma, the above formulas can be put together to a computation of $\vartheta(\sigma)$ for any ϑ and chain σ.

18.3 A Calculation

In this section we are concerned with periods of the form $2 \log \sigma(u) - \zeta(u)u$. This is an incomplete period of the third kind. Such a period appears for the first time in transcendence theory and it is natural to ask whether it is transcendental. However, there is no way to answer this directly; it has to be isolated from a linear form in incomplete elliptic periods. To achieve this we definitely need our results about linear independence of periods given in Chapter 11.

Proposition 18.3. *Let $P = \exp_E(u)$ and $Q = \exp_E(w)$ be distinct and non-zero points on E and let $\delta: [0,1] \to \mathbb{C}$ be a path from $-w$ to w such that $\gamma = \exp_E \circ \delta$ does not pass through P. Then*

$$\int_{-Q}^{Q} \xi_P := \int_\gamma \xi_P = 2 \log \frac{\sigma(u)\sigma(w)}{\sigma(w+u)} + 2\zeta(u)w + \log(-\wp(w) + \wp(u)) + 2\pi i v$$

for some $v \in \mathbb{Z}$. In the case $w = -u/2$, we write $P/2 = \exp_E(u/2)$ and then

$$\int_{P/2}^{-P/2} \xi_P := \int_\gamma \xi_P = 2 \log \sigma(u) - \zeta(u)u + \log\left(-\wp\left(\frac{u}{2}\right) + \wp(u)\right) + 2\pi i v.$$

Proof Incomplete periods of elliptic integrals are up to integer multiples of $2\pi i$ of the form $\log F(w,u)/F(v,u)$. We specialise to $v = -w$. Going back to the definition of $F(w,u)$, we calculate

$$
\begin{aligned}
\frac{F(w,u)}{F(-w,u)} &= \frac{\sigma(w-u)}{\sigma(-w-u)}\frac{\sigma(-w)}{\sigma(w)}e^{2\zeta(u)w} \\
&= \frac{\sigma(w-u)}{\sigma(w+u)}e^{2\zeta(u)w} \\
&= \frac{\sigma(w-u)\sigma(w+u)}{\sigma(w+u)^2}e^{2\zeta(u)w} \\
&= \frac{\sigma(u)^2\sigma(w)^2}{\sigma(w+u)^2}\left(-\wp(w)+\wp(u)\right)e^{2\zeta(u)w}
\end{aligned}
$$

using the identity

$$
\frac{\sigma(v+u)\sigma(v-u)}{\sigma(v)^2\sigma(u)^2} = -\wp(v)+\wp(u),
$$

which we take from (14) in [Fri11, p. 217]. This proves our first formula.

We continue with the choice $w = -u/2$ to get

$$
\frac{F(-\frac{u}{2},u)}{F(\frac{u}{2},u)} = \sigma(u)^2 e^{-\zeta(u)u}\left(-\wp(\tfrac{u}{2})\right)+\wp(u)),
$$

proving the second formula. $\qquad\square$

In the case of interest for us, E is defined over $\overline{\mathbb{Q}}$, and the points P and Q are chosen in $E(\overline{\mathbb{Q}})$. Then $\wp(u)$ and $\wp(u/2)$ are algebraic and hence $\int_{-P/2}^{P/2}\xi_P$ is equal to $2\log\sigma(u) - \zeta(u)u$ modulo Baker periods and multiples of $2\pi i$.

18.4 Transcendence of Incomplete Periods

We now come back to Schneider's Problem 3 mentioned in the introduction; see [Sch57, p. 138].

Theorem 18.4. *Let E be an elliptic curve over $\overline{\mathbb{Q}}$, P a point in $E(\overline{\mathbb{Q}})$ and ω, η, ξ_P the differential forms of the first, second and third kind from above. Assume that P is non-torsion in $E(\overline{\mathbb{Q}})$. Let $\kappa = \sum_{i=1}^{n}a_i\gamma_i$ be a chain in E^{an} with boundary in $E(\overline{\mathbb{Q}})$ avoiding the points e_∞, and P be such that $P(\kappa) = \sum_{i=1}^{n}a_i(\gamma_i(1)-\gamma_i(0)) \in E(\overline{\mathbb{Q}})$ is not a torsion point. Then*

$$
1, \quad 2\pi i, \quad \omega(\kappa), \quad \eta(\kappa), \quad \xi_P(\kappa)
$$

are $\overline{\mathbb{Q}}$-linearly independent, in particular $2\pi i, \omega(\kappa), \eta(\kappa), \xi_P(\kappa)$ are transcendental.

Proof We define $E^\circ = E \setminus \{e_\infty, P\}$. Let $D \subset E$ be the support of the boundary $\partial \kappa$. The chain κ defines a homology class $[\kappa] \in H_1^{\mathrm{sing}}(E^{\circ,\mathrm{an}}, D; \mathbb{Q})$, and the forms ω, η, ξ_P define classes in $H_{\mathrm{dR}}^1(E^\circ, D)$. This shows that we may view our periods as cohomological periods for $H^1(E^\circ, D)$ in the sense of Definition 12.2.

We choose an embedding $v^\circ \colon E^\circ \to J(E^\circ)$ into the generalised Jacobian $J(E^\circ)$ introduced in Section 4.5 via a base point $P_0 \neq e_\infty, P$. This is an extension of E by $\mathbb{G}_m$. Our assumption on P ensures that it is non-split up to isogeny. The induced map $v \colon E \to J(E) \cong E$ is $Q \mapsto Q + P_0$.

By Lemma 12.9, the periods of $H^1(E^\circ, D)$ agree with the periods of the 1-motive $[\mathbb{Z}[D]^0 \to J(E^\circ)]$. In fact, the submotive $M = [\mathbb{Z} \to J(E^\circ)]$ with $1 \mapsto P^\circ(\kappa) := \sum_i a_i v^\circ(\gamma_i(1) - \gamma_i(0))$ suffices. Note that the image of $P^\circ(\kappa)$ in $E(\overline{\mathbb{Q}})$ is $P(\kappa)$, which is independent of the choice of P_0. By assumption, it is non-torsion. Hence M is reduced as a 1-motive (see Definition 15.1) and of the form considered in Chapter 11. The class $[\kappa]$ can be identified with an element $\lambda \in V_{\mathrm{sing}}(M)$ as in the proof of Proposition 11.1, the non-CM-case, or Proposition 11.3, the CM-case. Accordingly, the periods of λ agree with the period integral of κ. As a consequence, the elements $1, 2\pi i, \omega(\kappa), \eta(\kappa), \xi_P(\kappa)$ are a subset of the basis considered there; in particular they are linearly independent. □

Remarks 18.5. 1. The assumption is satisfied if κ is a single non-closed path γ with $\gamma(1) - \gamma(0)$ non-torsion. In this case, the period numbers were computed explicitly in Summary 18.1 in terms of the Weierstraß functions. The cases of integrals of the first and second kind are actually already due to Schneider; see [Sch57, Satz 15, p. 60]. The result is new for integrals of the third kind.

2. It is possible to extend the considerations to the case where $P(\kappa)$ is torsion. We do not go into details here.

By specialising further, we obtain the following explicit transcendence result.

Theorem 18.6. *Let $u \in \mathbb{C}$ be such that $\wp(u) \in \overline{\mathbb{Q}}$ and $\exp_E(u)$ is non-torsion in $E(\overline{\mathbb{Q}})$. Then*

$$u\zeta(u) - 2\log\sigma(u)$$

is transcendental.

Proof For $\exp_E(u)$ we simply write P. We choose a path δ from $u/2$ to $-u/2$ which avoids the singularities of ξ_P and put $\gamma = \exp_E \circ \delta$. By Proposition 18.3, the period has the form

$$\xi_P(\gamma) = 2\log\sigma(u) - \zeta(u)u + \log(\alpha) + 2\pi i v$$

for some algebraic α. By Theorem 18.4, it is transcendental, but this is not enough. If

$$2\log\sigma(u) - \zeta(u)u = \xi_P(u) - \log\alpha - 2\pi i v$$

was algebraic, then we would have a linear dependence relation between the numbers $1, \xi_P(\gamma), \log\alpha, 2\pi i$. To obtain a contradiction, it suffices to show that they are $\overline{\mathbb{Q}}$-linearly independent. (Except when α is a root of unity and $\log\alpha$ a rational multiple of $2\pi i$. Then the element $\log\alpha$ can be dropped from the list and the linear independence is already shown in Theorem 18.4.)

Note that the term $\log\alpha$ is the period of a Kummer motive $M_0 = [\mathbb{Z} \to \mathbb{G}_m]$ with $1 \mapsto \alpha$ and $\xi_P(\gamma)$ is an incomplete period of the third kind of $M_1 = [\mathbb{Z} \to J(E^\circ)]$, as in the proof of Theorem 18.4.

Linear independence could be addressed by applying the techniques of Chapter 15 directly to $M_0 \times M_1$. Instead we explain the deduction from the general results proved earlier. In fact, as shown in Example 17.16, we have $\delta_{\mathrm{inc}3}(M_1) = 1$. This means that $\xi_P(\gamma)$ is a non-zero element of $\mathcal{P}_{\mathrm{inc}3}(M_1)$ (see Chapter 16). By Lemma 15.24, we have $\mathcal{P}_{\mathrm{inc}3}(M_1) \subset \mathcal{P}_{\mathrm{inc}3}(M_1^{\mathrm{sat}})$ for M_1^{sat} a saturation of M_1 as constructed there. Proposition 16.5, item (5) shows that $\log\alpha$ and $\xi_P(\gamma)$ are linearly independent in

$$\mathcal{P}_{\mathrm{inc}3}(M_0 \times M_1^{\mathrm{sat}}) = \mathcal{P}\langle M_0 \times M_1^{\mathrm{sat}}\rangle / (\mathcal{P}\langle \mathbb{G}_m \times J(E^\circ)\rangle + \mathcal{P}\langle [\mathbb{Z} \to E]\rangle).$$

The terms $1, 2\pi i$ are in $\mathcal{P}\langle \mathbb{G}_m \times J(E^\circ)\rangle + \mathcal{P}\langle [\mathbb{Z} \to E]\rangle$ and linearly independent by Lindemann's result; see Corollary 10.1 or Theorem 15.3. We deduce that all four are linearly independent and $u\zeta(u) - 2\log\sigma(u)$ is transcendental. $\square$

18.5 Elliptic Period Space

In [Wüs21], period spaces for elliptic curves and abelian varieties of dimension 2 were considered and their dimension was determined; see [Wüs21, Theorems 2 and 3, respectively]. The motivation was billiards on the ellipsoids where irrationality of elliptic and abelian periods give an answer to the question of whether curvature lines or geodesics are closed or not. The proofs rely on the Analytic Subgroup Theorem in its original version.

In this section we come back to this problem in a more general setting using the language of 1-motives. We confine ourselves to the elliptic case but we go further and determine the dimension of an extended period space: a period space where, instead of considering a single differential of the third kind we consider a finite number. In addition we do not restrict to closed paths as in

[Wüs21] but allow that the paths need not be closed. This was mentioned as a problem in [Wüs21]. It turns out that the presence of several differentials of the third kind and the additional generality of allowing non-closed paths makes the problem much more difficult and gives a rather unexpected answer for specialists.

Suppose now that E is an elliptic curve defined over $\overline{\mathbb{Q}}$, ω, η are the differentials of the first and the second kind and $\xi_k = \xi_{P_k}$, $1 \le k \le n$, are differentials of the third kind on E with $P_k = \exp_E(u_k) \in E(\overline{\mathbb{Q}})$ for $u_k \in \mathrm{Lie}(E)_{\mathbb{C}}$; see Section 18.1, equation (18.2). We denote by ω_1, ω_2, η_1 and η_2 the periods and quasi-periods of E in the classical sense, i.e. the integrals of ω and η with respect to a pair of basis vectors of $H_1(E^{\mathrm{an}}, \mathbb{Z})$. We choose non-closed paths $\gamma_i: [0,1] \to E^{\mathrm{an}}$, $1 \le i \le m$, with $\gamma_i(0), \gamma_i(1) \in E(\overline{\mathbb{Q}})$.

Definition 18.7. The period space $W = W(E, P_k, \gamma_i)$ is generated over $\overline{\mathbb{Q}}$ by

$$1,\ 2\pi i,\ \omega_1,\ \omega_2,\ \eta_1, \eta_2,\ \lambda(u_k, \omega_1),\ \lambda(u_k, \omega_2),\ \omega(\gamma_i),\ \eta(\gamma_i),\ \xi_{P_k}(\gamma_i)$$

for $1 \le i \le m$ and $1 \le k \le n$.

We will show that W can be identified with the period space of a 1-motive. Let S be the set $S = \{0, P_1, \ldots, P_n\} \subset E$ and D the union of the supports of $\partial \gamma_i$ for $i = 1, \ldots, m$. Put $E^\circ = E \smallsetminus S$. Consider the object $H^1(E^\circ, D)$ in the category $(\overline{\mathbb{Q}}, \mathbb{Q})$-Vect. By Lemma 12.9, its periods agree with the periods of the 1-motive

$$M' = [\mathbb{Z}[D]^0 \to J(E^\circ)].$$

Let $L \subset \mathbb{Z}[D]^0$ be generated by $\partial \gamma_i$ for $i = 1, \ldots, m$ and put

$$M = [L \to J(E^\circ)].$$

The lattice L has rank at most m. We write $G = J(E^\circ)$ for short. It has abelian part E and a torus part T of rank at most n.

Proposition 18.8. *The period space W coincides with $\mathcal{P}\langle M \rangle$.*

Proof We choose generators for

$$V_{\mathrm{sing}}(M) \subset V_{\mathrm{sing}}(M') \cong H_1^{\mathrm{sing}}(E^{\circ,\mathrm{an}}, D; \mathbb{Q})$$

in the following way. Take small loops $\sigma_0, \sigma_1, \ldots, \sigma_n$ around the points in S. They generate $V_{\mathrm{sing}}([0 \to T])$. Next choose loops $\varepsilon_1, \varepsilon_2$ in $E^{\circ,\mathrm{an}}$ whose images in E^{an} generate $H_1^{\mathrm{sing}}(E^{\mathrm{an}}, \mathbb{Z})$. Finally, view $\gamma_1, \ldots, \gamma_m$ as paths in $E^{\circ,\mathrm{an}}$. By definition,

$$\sigma_0,\ \ldots,\ \sigma_n,\ \varepsilon_1,\ \varepsilon_2,\ \gamma_1,\ \ldots,\ \gamma_m$$

generate $V_{\mathrm{sing}}(M)$.

We turn to $V_{\mathrm{dR}}^{\vee}(M)$, a quotient of $H_{\mathrm{dR}}^1(E^{\circ}, D)$. We take exact differential forms $u_1, \ldots, u_r$ which generate the kernel of $H_{\mathrm{dR}}^1(E^{\circ}, D) \to H_{\mathrm{dR}}^1(E^{\circ})$. The differential forms $\omega, \eta, \xi_{P_1}, \ldots, \xi_{P_n}$ can be seen as elements of $H_{\mathrm{dR}}^1(E^{\circ}, D)$. As we shall prove, the set $u_1, \ldots, u_m, \omega, \eta, \xi_{P_1}, \ldots, \xi_{P_n}$ generates the whole cohomology. Obviously $u_1, \ldots, u_m, \omega, \eta$ generate the subspace $H_{\mathrm{dR}}^1(E, D)$, and hence it remains to show that the ξ_{P_i} generate

$$H_{\mathrm{dR}}^1(E^{\circ}, D)/H_{\mathrm{dR}}^1(E, D) \cong H_{\mathrm{dR}}^1(E^{\circ})/H_{\mathrm{dR}}^1(E) \cong \ker(H_{\mathrm{dR}}^0(S) \to H_{\mathrm{dR}}^0(E)).$$

The composition of the two isomorphisms maps a logarithmics form ϑ with polar divisor included in S to its residue vector $(\mathrm{res}_0 \vartheta, \mathrm{res}_{P_1} \vartheta, \ldots, \mathrm{res}_{P_n} \vartheta)$. The image of ξ_{P_i} is $(-2, 0, \ldots, 2, 0, \ldots)$ with 2 in place i. Together they generate the kernel as claimed.

As a consequence, the period matrix of M has the shape

$$\begin{pmatrix} \xi_{P_i}(\sigma_j) & \xi_{P_i}(\varepsilon_j) & \xi_{P_i}(\gamma_i) \\ 0 & \omega(\varepsilon_i) & \omega(\gamma_j) \\ 0 & \eta(\varepsilon_i) & \eta(\gamma_j) \\ 0 & 0 & u_i(\gamma_j) \end{pmatrix} = \begin{pmatrix} 2\pi i \alpha_{ij} & \lambda(u_i, \omega_j) + 2\pi v_j & \xi_{P_i}(\gamma_i) \\ 0 & \omega_i & \omega(\gamma_j) \\ 0 & \eta_i & \eta(\gamma_j) \\ 0 & 0 & \beta_{ij} \end{pmatrix},$$

with $\alpha_{ij}, \beta_{ij} \in \overline{\mathbb{Q}}$, $v_j \in \mathbb{Z}$. Its entries generate the vector space W.   □

Corollary 16.4 together with Proposition 16.5 give a formula for the dimension of W. With the notation introduced earlier, we state the following result.

Theorem 18.9. *Let $E/\overline{\mathbb{Q}}$ be elliptic with $e = \dim_{\mathbb{Q}} \mathrm{End}(E)_{\mathbb{Q}}$. The dimension of W over $\overline{\mathbb{Q}}$ is given by*

$$\dim_{\overline{\mathbb{Q}}} W = 2 + \frac{4}{e} + 2\mathrm{rk}_E(T, M) + 2\mathrm{rk}_E(L, M) + \delta_{\mathrm{inc}3}(M).$$

If M is saturated, then $\delta_{\mathrm{inc}3}(M) = e \cdot \mathrm{rk}_E(T, M) \cdot \mathrm{rk}_E(L, M)$.

Note that M is *not* reduced in general. If $L \to E(\overline{\mathbb{Q}})_{\mathbb{Q}}$ or $X(T) \to E(\overline{\mathbb{Q}})_{\mathbb{Q}}$ has a kernel, this implies that suitable Baker periods (i.e. values of log in algebraic numbers) are contained in W. This happens, for example, if the endpoint of a path or one of the P_k are torsion points in $E(\overline{\mathbb{Q}})$. The situation simplifies if we exclude this case.

Corollary 18.10. *Assume that E does not have CM, that $n = \mathrm{rk}\langle P_1, \ldots, P_n \rangle$ and that $m = \mathrm{rk}\langle \gamma_i(1) - \gamma_i(0) | i = 1, \ldots, m \rangle$ as subgroups of $E(\overline{\mathbb{Q}})$. Then*

$$\dim W = 6 + 2(n + m) + nm.$$

Proof The assumptions imply that $e(E) = 1$, that M is saturated and that $n = \mathrm{rk}_E(T, M)$, $m = \mathrm{rk}_E(L, M)$.   □

18.5.1 With CM

The CM-case is a lot more complicated, even if M is reduced. We consider an example with small rank $n = m = 2$, hence $M = [L \to G]$ is reduced, with $L \cong \mathbb{Z}^2$ and G an extension of an elliptic curve E by the torus $T = \mathbb{G}_m^2$ characterised by its classifying map $X(T) \to E^\vee$.

Assume that $K = \mathrm{End}_\mathbb{Q}(E)$ is an imaginary quadratic field, which gives $e = 2$. In this situation $\mathrm{rk}_E(L, M)$ and $\mathrm{rk}_E(T, M)$ can take the values 1 and 2 and then we can state the following computation.

Lemma 18.11.

$$\delta_{\mathrm{inc3}}(M) = \begin{cases} 2 & M \text{ saturated,} \\ 4 & \text{otherwise.} \end{cases}$$

Proof We show that if there is a non-zero element in $R_{\mathrm{inc3}}(M)$ then M is saturated. We go back to the characterisation of Theorem 17.15 and choose a $\mathbb{Z}$-basis l_1, l_2 of $L \subset E(\overline{\mathbb{Q}})$ which is used to make the identification $L^\vee \otimes E \cong E^2$, as before. This gives $c = (l_1, l_2)$. Let (α, y, x) be a triple satisfying the conditions in Theorem 17.15 with $y \neq 0$. First consider the morphism $\alpha \colon E^2 \to E^2$. We may view it as an element of $M_2(K)$. Condition (B) implies that $\alpha(c) = \alpha(l_1, l_2) = 0$. There are three possible cases, depending on the rank of α.

If α is invertible, then the non-zero vector (l_1, l_2) cannot be mapped to 0. This case does not occur.

If α has rank 1, it suffices to consider $\alpha' \colon E^2 \to \alpha(E) \cong E$. We replace α by α' in the arguments and x, y by their images in E and $X(T)$. The new α has shape (m, n) for $m, n \in K$. By assumption, the image of c in E is

$$\alpha'(c) = m\alpha'(l_1) + n\alpha'(l_2) = ml_1' + nl_2' = 0.$$

Without loss of generality, $m \neq 0$, hence it is invertible. We replace α by $m^{-1} \circ \alpha$ and then have $m = 1$, $l_1' + nl_2' = 0$ with n replaced by $m^{-1}n$. As l_1' and l_2' are $\mathbb{Z}$-linearly independent, this implies $n \in K \setminus \mathbb{Q}$. The image of L in $E(\overline{\mathbb{Q}})$ contains l_2' and $-nl_1'$. As $1, -n$ are a $\mathbb{Q}$-basis of K, this implies $Kl_2' \subset L$, and since L has $\mathbb{Q}$-rank 2, this even implies $L \cong Kl_2'$. In other words, the lattice L is K-stable in $E(\overline{\mathbb{Q}})_\mathbb{Q}$ and we have $\mathrm{rk}_E(L, M) = 1$.

We continue with the diagram (A). As M is reduced, the classifying map $X(T) \to E^\vee(\overline{\mathbb{Q}})$ is injective and we use it to identify elements of $X(T)$ with elements of $E^\vee(\overline{\mathbb{Q}})$. The adjoint of α is of the form $(1, n^\vee) \colon E^\vee \to E^{\vee 2}$. This gives $y = \alpha^\vee(x) = (x, n^\vee(x)) \in X(T)^2$. In particular, x is a non-zero element of $X(T)$ and its image under $n^\vee \in K$ is again in $X(T)$. Hence $X(T)$ is also K-stable in $E^\vee(\overline{\mathbb{Q}})$ and $\mathrm{rk}_E(T, M) = 1$. This implies that the classifying map of G is K-equivariant and that the operation of K on E extends to an action on G.

The map in (A) is $G^2 \to G$, given by $(1, n)$. Condition (B) is $l_1 + nl_2 = 0$, and if this is satisfied, then $L \subset G(\overline{\mathbb{Q}})$ is K-stable, which means that M is saturated. In this case,

$$\delta_{\text{inc}3}(M) = e\text{rk}_E(L, M)\text{rk}_E(T, M) = 2.$$

It remains to consider the case $\alpha = 0$, in which the condition $[L^\vee \otimes G]^*(y) = \alpha^*(x) = 0$ implies $y = 0$ and then $\delta_{\text{inc}3}(M) = 4$. $\qquad\qquad\qquad\qquad\square$

Corollary 18.12. *The possible values for $\delta(M)$ in the CM-case are 16 when neither neither L nor $X(T)$ are K-stable, 14 if one of L and $X(T)$ is K-stable, 12 if both are K-stable but M is not saturated, and 10 if M is saturated.*

19

Values of Hypergeometric Functions

We review how our knowledge of periods of curves and 1-motives can also be used to deduce transcendence results for certain values of hypergeometric functions. The result in the elliptic case can be found as a special case of Wolfart's publication [Wol88] or in [CC88] by Chudnovsky–Chudnovsky.

More generally it is well known that values of hypergeometric functions can be expressed as a quotient of two abelian integrals, in general of the second kind. This leads to a period relation between the two periods of the second kind with the hypergeometric function as coefficient. Algebraic values of the hypergeometric function provide linear relations between the two periods with algebraic coefficients. This cannot be true in general and leads to special points on certain Shimura varieties as explained very carefully in Chapter 5 of Tretkoff's beautiful monograph [Tre17].

19.1 Elliptic Integrals

We fix a parameter $\lambda \in \mathbb{C} \setminus \{0, 1\}$. The famous differential form

$$\xi(\lambda) = \frac{du}{\sqrt{u(1-u)(1-u\lambda)}}$$

on the compactification $\hat{\mathbb{C}}$ of the complex plane is multi-valued with branch points $0, 1, \infty, \lambda^{-1}$, the so-called Weierstraß points. Locally, the branches differ by a sign.

Integration of differential forms of the type $\xi(\lambda)$ over paths leads to so-called Euler integrals, which were introduced and studied by Euler in connection with his work on hypergeometric functions.

In our special case of the differential form $\xi(\lambda)$, we take the integrals over arcs $\gamma_{p,q}$ with loose end at two of the four branch points p and q, respectively, but not passing through branch points. They are given by the improper integrals

$$I_{p,q}(\lambda) = \int_{\gamma_{p,q}} \frac{1}{\sqrt{u(1-u)(1-\lambda u)}} \, du,$$

and there are six choices of pairs. Each choice of such a path determines the value up to sign. The convergence of these integrals can be seen as follows: we take the path $\gamma_{0,1}$ and have to show that the integral converges locally at 0 and 1. The change of variables $t^2 = u$ gives convergence close to 0, and the change of variables $t^2 = 1 - u$ gives convergence close to 1. The remaining integrals can be transformed by applying the Moebius transformations

$$u \mapsto \frac{1}{u}, \quad u - 1, \quad \frac{1}{u-1}, \quad \frac{(u-1)}{u}, \quad \frac{u}{1-u}.$$

Example 19.1. If $\lambda \notin [1, \infty)$, we choose $\gamma_{0,1}$ as the straight path from 0 to 1 in $\mathbb{C}$. By definition, the function u^a for complex $a \in \mathbb{C}$ is given by $\exp(a(\log|u| + i\arg u))$. We take the branch of the integrand determined by the following assignment of the arguments,

$$\arg u = 0, \quad \arg(1-u) = 0, \quad |\arg(1-\lambda u)| < \pi/2,$$

and then, as we have indicated above, the integral is convergent.

The Euler integral can be viewed as a period in our sense. To see this, let C_λ be the curve of genus 1 in Legendre form with affine equation

$$y^2 = u(1-u)(1-\lambda u).$$

Since the point ∞ is rational, the curve has a group structure and becomes an elliptic curve.

The projection $\pi \colon C_\lambda \to \mathbb{P}^1$ with $(u, y) \mapsto u$ is a 2-fold cover ramified in the 2-torsion points $u = 0, 1, \lambda^{-1}, \infty$. The multi-valued differential form $\xi(\lambda)$ lifts from $\hat{\mathbb{C}}$ to the single-valued form and we have

$$\omega(\lambda) = \frac{du}{y}$$

on C_λ. The closure of our arcs $\gamma_{p,q}$ in $\mathbb{P}^1$ lift to paths $\widetilde{\gamma}_{p,q}$ on C_λ^{an} and then

$$I_{p,q}(\lambda) = \int_{\widetilde{\gamma}_{p,q}} \omega(\lambda).$$

There are two choices for the lift, and the choice of the branch over $\gamma_{p,q}$ in the original definition is replaced by the choice of the lift $\widetilde{\gamma}_{p,q}$. Let $\widetilde{\gamma}_{p,q}^-$ be the other lift. Then $[\widetilde{\gamma}_{p,q}^-] = -[\widetilde{\gamma}_{p,q}]$. For algebraic λ, this description makes $I_{p,q}(\lambda)$ an incomplete elliptic period of the first kind corresponding to a motive of the form $[\mathbb{Z} \to C_\lambda]$. This insight is enough for the application to values of the hypergeometric function, but we can be more precise. As the endpoints are 2-torsion points, the image of $\widetilde{\gamma}_{p,q}$ is a closed path in $C_\lambda/C_\lambda[2]$. The elliptic integral $I_{p,q}(\lambda)$ agrees with a complete elliptic period of the first kind there. Expressed in terms of 1-motives: $[\mathbb{Z} \to C_\lambda]$ is isomorphic to $[\mathbb{Z} \to 0] \oplus [0 \to C_\lambda]$ in the isogeny category 1-Mot$_{\overline{\mathbb{Q}}}$ and the integral $I_{p,q}(\lambda)$ is a period of the second factor.

Another way to see this is by complex analysis. The standard way in complex analysis is to replace the loose ends by small circles with radius ϵ. We get a closed path $c_{p,q}$ around p and q by going back and forth along the arc from p to q and around the circles.

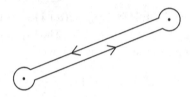

Since the differential form is multi-valued, one has to be careful: going once around the circle counterclockwise changes the value of $\xi(\lambda)$ by -1 along the arc. Going once around the second circle again counterclockwise does the same and we get back the original determination of the value of the differential form along the arc but we are passing in the opposite direction. The value of the integral is independent of ε. After letting ϵ tend to zero, we obtain $\int_{c_{p,q}} \xi(\lambda) = 2\,I_{p,q}(\lambda)$. Our paths $c_{p,q}$ lift to closed paths $\widetilde{c}_{p,q}$ on C_λ^{an} such that

$$[\widetilde{c}_{p,q}] = [\widetilde{\gamma}_{p,q}] - [\widetilde{\gamma}_{p,q}^-] = 2[\widetilde{\gamma}_{p,q}] \in H_1(C_\lambda^{\mathrm{an}}, \mathbb{Q})$$

and

$$2\,I_{p,q}(\lambda) = \int_{\widetilde{c}_{p,q}} \omega(\lambda).$$

For algebraic λ, this description makes $I_{p,q}(\lambda)$ a complete elliptic period of the first kind corresponding.

In toto:

Lemma 19.2. *Let p, q, r be distinct in $\{0, 1, \infty, \lambda^{-1}\}$. Then $\widetilde{c}_{p,q}$ and $\widetilde{c}_{p,r}$ form a basis of $H_1^{\mathrm{sing}}(C_\lambda^{\mathrm{an}}, \mathbb{Q})$.*

Proof As a topological space we may identify C_λ^{an} with $(\mathbb{R}/2\mathbb{Z})^2$. Our exceptional points are the classes of $(0,0)$, $(1,0)$, $(0,1)$, $(1,1)$. We lift the paths $\gamma_{p,q}$ to the universal cover $\mathbb{R}^2$ where they start in a lift $\widetilde{p}$ of p and end in lifts $\widetilde{q}$ and $\widetilde{r}$ of q and r, respectively. The alternative lift $\widetilde{\gamma}_{p,q}^-$ connects $-\widetilde{p}$ to $-\widetilde{q}$. Up to homotopy the lift of the closed loop $c_{p,q}$ connects $\widetilde{p}$ to $\widetilde{p} + 2(\widetilde{q} - \widetilde{p})$. Its homology class is $2(\widetilde{q} - \widetilde{p}) \in (2\mathbb{Z})^2 = H_1^{\mathrm{sing}}((\mathbb{R}/2\mathbb{Z})^2, \mathbb{Z})$. The vectors $\widetilde{q} - \widetilde{p}, \widetilde{r} - \widetilde{p}$ are non-zero and distinct in $(\mathbb{Z}/2\mathbb{Z})^2$, hence linear independent. This makes $2(\widetilde{q} - \widetilde{p}), 2(\widetilde{r} - \widetilde{p})$ linear independent in $(2\mathbb{Z})^2$. $\qquad\qquad\square$

19.1.1 A Hypergeometric Function

The hypergeometric function with parameters $(1/2, 1/2, 1)$ is defined by the power series

$$F\left(\frac{1}{2}, \frac{1}{2}, 1; \lambda\right) = \sum_{n=0}^{\infty} \frac{(\frac{1}{2})_n (\frac{1}{2})_n}{n!} \frac{\lambda^n}{n!},$$

which is convergent for $|\lambda| < 1$. Here $(a)_n = a(a+1) + \cdots + (a+n-1)$ for $n > 0$, and 1 for $n = 0$, are the Pochhammer symbols. The hypergeometric function is a solution of the hypergeometric differential equation

$$\lambda(\lambda - 1)\phi'' + (2\lambda - 1)\phi' + \frac{1}{4}\phi = 0. \tag{19.1}$$

As such it extends to a meromorphic function on $\mathbb{C} \smallsetminus \{0, 1\}$. The differential equation is a second-order equation of Fuchsian type with regular singular points at $0, 1, \infty$. Its solution space W has dimension 2. A basis for W is given by the pair

$$z_0(\lambda) := \quad F\left(\frac{1}{2}, \frac{1}{2}, 1; \lambda\right),$$

$$z_1(\lambda) := -iF\left(\frac{1}{2}, \frac{1}{2}, 1; 1 - \lambda\right).$$

Each solution extends to a meromorphic function on $\mathbb{C} \smallsetminus \{0, 1\}$.

A classic computation due to Euler (see [Kle81, p. 7]; see also [IKSY91, Section 2.3.2]) shows that for $|\lambda| < 1$,

$$I_{0,1}(\lambda) = \mathsf{B}\left(\frac{1}{2}, \frac{1}{2}\right) F\left(\frac{1}{2}, \frac{1}{2}, 1; \lambda\right), \tag{19.2}$$

where $\mathsf{B}(p, q)$ is Euler's Beta-function. From

$$\mathsf{B}\left(\frac{1}{2}, \frac{1}{2}\right) = \int_0^1 u^{-1/2}(1 - u)^{-1/2} du = \frac{\Gamma(\frac{1}{2})\Gamma(\frac{1}{2})}{\Gamma(1)} = \pi,$$

we conclude that for all $\lambda \in \mathbb{C} \smallsetminus \{0, 1\}$,

$$I_{0,1}(\lambda) = \pi F\left(\frac{1}{2}, \frac{1}{2}, 1; \lambda\right). \tag{19.3}$$

We know that π and $2I_{0,1}(\lambda)$ are periods. For π this is clear and for $I_{0,1}(\lambda)$ we discussed it at length in Section 19.1. Summary 14.11 implies that $I_{0,1}(\lambda)$ is non-zero if λ is algebraic. Equation (19.3) is a $\mathbb{C}$-linear relation between the elliptic period $I_{0,1}(\lambda)$ and π, which, by Theorem 16.2, are linearly independent over $\overline{\mathbb{Q}}$. These considerations prove the following result.

Proposition 19.3 (Wolfart, Chudnovsky–Chudnovsky). *For $z \in \overline{\mathbb{Q}} \smallsetminus \{0, 1\}$, the value $F(1/2, 1/2, 1; z)$ of the hypergeometric function is transcendental.*

Remark 19.4. The above fact is pointed out by Wolfart [Wol88, §3, Fall 3] as a consequence of [WW85, Satz 2]. Chudnovsky–Chudnovsky mention it in [CC88, p. 426] as a corollary of Chudnovsky's Theorem on the algebraic independence of elliptic periods. Both references study more generally values of hypergeometric functions from different angles. André [And96] has an alternative approach. But, in fact, the necessary transcendence result is much older: Schneider [Sch37, Satz IIIa] proved in 1936 the transcendence of π/ω where ω is a period of an elliptic curve defined over $\overline{\mathbb{Q}}$. As the relation to values of the hypergeometric function that we described is classical, we do not know who was the first to make the connection to their transcendence.

An alternative 'modular' proof is suggested by the explicit computation of $F(1/2/, 1/2, 1; z)$ given in [Arc03a, Remark 6], corrected by Bostan; see [Bos21]. We have

$$F\left(\frac{1}{2}, \frac{1}{2}, 1; \lambda(\tau)\right) = E_4(\tau)^{\frac{1}{4}}(\lambda^2(\tau) - \lambda(\tau) + 1)^{-\frac{1}{4}}, \qquad (19.4)$$

where $\lambda(\tau)$ is the Legendre modular function and $E_4(\tau) = (3/4\pi^4)\,g_2(\tau)$ the Eisenstein modular form of weight 4. The difference to Archinard's formula is the inverse on the left-hand side. Assume $\lambda(\tau)$ and $F(1/2, 1/2, 1; \lambda(\tau))$ to be algebraic. This makes $j(\tau)$ and $E_4(\tau)$ algebraic. The identity

$$j(\tau) = 1728\frac{E_4(\tau)^3}{E_4(\tau)^3 - E_6(\tau)^2}$$

implies that $E_6(\tau)$ is algebraic as well. This is a contradiction to what Bertrand [Ber76, §1.1] showed. As a consequence, $F(1/2, 1/2, 1; \lambda(\tau))$ is transcendental for algebraic $\lambda(\tau)$. We thank Bostan for pointing out this argument. Looking more closely, we see that this is actually the same as before: Bertrand's proof relies on the $\overline{\mathbb{Q}}$-linear dependence between π and elliptic periods.

19.1.2 The Legendre Family

We return to the integrals $I_{p,q}(\lambda)$ as functions in the variable $\lambda \in S^{\mathrm{an}} :=$ $\hat{\mathbb{C}} \smallsetminus \{0, 1, \infty\}$. The functions are holomorphic and multi-valued. We concentrate on $I_{0,1}(\lambda)$. Different branches correspond to different choices of homotopy classes of paths $\gamma_{0,1}$.

Example 19.5. On $\hat{\mathbb{C}} \smallsetminus [1, \infty)$, we use the principal branch of the function normalised as in Example 19.1. It does not extend to $\lambda \in [1, \infty)$ because the straight path $[0, 1]$ would pass through the branch point λ^{-1} and we would get two possible values for the integral depending on the choice of the branch. Instead, choose $\lambda_0 \in [1, \infty)$ and replace the straight path via one of the two semi-circles with sufficiently small radius around λ_0^{-1}.

The choice of the semi-circle replaces the choice of branch and we take the one on which $\Im(z) > 0$. This point of view gives a description of the analytic continuation of $I_{0,1}(\lambda)$. Let γ be the new path. The integral $\int_\gamma \omega(\lambda)$ is well defined for $\lambda \in \mathbb{C} \smallsetminus \{\gamma(t)^{-1} \mid t \in [0,1]\}$, as can be verified, and defines a holomorphic function. In particular, it is well defined at λ_0. By the Monodromy Theorem, the modified function agrees with $I_{0,1}(\lambda)$ for λ^{-1} in the lower half plane because $\omega(\lambda)$ remains regular between the two paths. It furnishes an analytic continuation of $I_{0,1}(\lambda)$, depending on our choice of γ. If instead we take the second semi-circle, we obtain a second analytic continuation. Going through all possible paths from 0 to 1 gives the full analytic continuation of $I_{0,1}(\lambda)$ as a multi-valued function on $\mathbb{C} \smallsetminus \{0, 1\}$. In particular, all values of the analytic continuation are periods for different choices of path from 0 to 1.

The Euler integrals are solutions of the hypergeometric equation

$$\lambda(\lambda - 1)\phi'' + (2\lambda - 1)\phi' + \frac{1}{4}\phi = 0. \tag{19.5}$$

We refer the reader to [Kle81, §16] or [IKSY91] for the explicit computation. This is a second-order differential equation of Fuchsian type with regular singular points at $0, 1, \infty$. The fundamental group of $S^{an} = \mathbb{P}^1 \smallsetminus \{0, 1, \infty\}$ is a free group Γ with generators γ_0, γ_1 and γ_∞, which are loops with base point $b = 1/2 \in S^{an}$ around the points $0, 1, \infty$ with $\gamma_0^{-1} = \gamma_1\gamma_0$. The solution space W of the differential equation is a local system of rank 2. The group Γ has a representation ρ, the monodromy representation, in the solution space W. Accordingly the solutions are multi-valued functions on S^{an} and holomorphic on the universal cover.

A more conceptual interpretation of the situation is to consider the Legendre family

$$p \colon C \to \mathbb{P}^1$$

with fibre C_λ at λ. The family has degenerate fibres at 0, 1, ∞. Over $S = \mathbb{P}^1 \setminus \{0, 1, \infty\}$ there is a global basis of $H^1_{dR}(C_\lambda)$ given by the two differential forms

$$\omega(\lambda) = \frac{du}{y}, \quad \eta(\lambda) = \frac{udu}{y}.$$

The form $\omega(\lambda)$ is holomorphic and thus of the first kind, whereas $\eta(\lambda)$ has a pole of order 2 at $u = \infty$ and this means that it is of the second kind.

The homology groups $H^{sing}_1(C^{an}_\lambda, \mathbb{Z})$ organise as a local system of rank 2 on S^{an}. Abstractly, it is the dual of $p_{S*}\mathbb{Z}$, where p_S is the Legendre family over S. For an explicit description let $s \in S^{an}$ be fixed and let γ_s be a cycle in C_s. By parallel transport, we obtain a horizontal lifting $\gamma(\lambda)$ of the cycle γ_s. It depends on the choice of an Ehresmann connection. But the homology class of the cycle $\gamma(\lambda)$ is independent of the choice. This gives a horizontal family of homology classes of cycles. The periods

$$z(\lambda) := \int_{\gamma(\lambda)} \omega(\lambda)$$

along horizontal cycles give multi-valued analytic functions on S. They are solutions of the Gauß–Manin connection on $H^1_{dR}(C_\lambda/S)$, which leads to a differential equation with regular singularities in $\{0, 1, \infty\}$. The differential equation so obtained coincides with the differential equation (19.5).

19.2 Abelian Integrals

In the previous section we introduced the special hypergeometric function $F(1/2, 1/2, 1; \lambda)$ and showed that for $\lambda \neq 0$ it takes transcendental values. Our proof is based on the comparison of the periods of the Legendre curve and the period of the curve $y^2 = x(1 - x)$ obtained from the Legendre curve when $\lambda = 0$. This will now be studied for superelliptic generalisations of the Legendre curve.

19.2.1 Euler Integral

The hypergeometric function $F(1/2, 1/2, 1; \lambda)$ is only a very particular case of the hypergeometric function $F(a, b, c; \lambda)$ with expansion

$$F(a, b, c; \lambda) = \sum_{n=0}^{\infty} \frac{(a)_n (b)_n}{(c)_n} \frac{\lambda^n}{n!}$$

and convergent for $|\lambda| < 1$. As before, $(a)_n = a(a + 1) \ldots (a + n - 1)$ are the Pochhammer symbols. In the most general case, the arguments a, b and c are complex numbers with c neither zero nor a negative integer. The hypergeometric function $F(a,b,c;\lambda)$ satisfies the differential equation

$$\lambda(1-\lambda)F'' + (c-(a+b+1)\lambda)F' - abF = 0.$$

One sees that the differential equation specialises to the differential equation (19.5) from above when taking $a = b = 1/2$. Also for this more general hypergeometric function an integral representation can be derived. We consider the differential forms

$$\omega(a,b,c;\lambda) = u^{b-1}(1-u)^{c-b-1}(1-\lambda u)^{-a}du \qquad (19.6)$$

and

$$\omega(b,c-b) = u^{b-1}(1-u)^{c-b-1}du.$$

The Euler integral

$$\Omega(a,b,c;\lambda) = \int_0^1 u^{b-1}(1-u)^{c-b-1}(1-\lambda u)^{-a}du$$

can be expressed in terms of $F(a,b,c;\lambda)$ and the Euler Beta-function. The latter is usually written as

$$\mathsf{B}(b,c-b) = \int_0^1 u^{b-1}(1-u)^{c-b-1}du.$$

It is obtained from the degeneration of $\omega(a,b,c;\lambda)$ at $\lambda = 0$.

Proposition 19.6 ([Kle81, p. 7], [IKSY91, Section 2.3.2])). *The Euler integral and the hypergeometric function $F(a,b,c;\lambda)$ are related by the equation*

$$\Omega(a,b,c;\lambda) = \mathsf{B}(b,c-b)F(a,b,c;\lambda).$$

If a, b and c are rational numbers with smallest common denominator N, then $\Omega(a,b,c;\lambda)$ can be interpreted as a period on the algebraic curve $C_N(\lambda)$ of the form

$$y^N = x^r(1-x)^s(1-\lambda x)^t$$

for suitable r,s,t. The degeneration $C_N(0)$ with affine equation

$$y^N = x^r(1-x)^s$$

has the same property with respect to $\mathsf{B}(b,c-b)$. As in the elliptic case, knowledge about linear independence of periods leads to transcendence results for $F(a,b,c;\lambda)$. This connection was already exploited by Wolfart in [Wol88]. We explain an approach from a different angle. In order to simplify the exposition, we restrict the discussion to the case where $N = p$ is an odd prime, $0 < r,s,t < p$, and require that p does not divide $r+s+t$ or $r+s$.

19.2.2 Geometry of $C_p(\lambda)$

Let $\lambda \neq 0, 1$ and p, r, s, t be as specified above. The curve $C_p(\lambda)$ was studied first by Wolfart, then in great detail by Archinard [Arc03b], who corrected some errors by Wolfart. Quite recently Archinard's paper was extended and corrected by Asakura and Otsubo [AO18]. We briefly sketch the results that are relevant for us.

The curves $C_p(\lambda)$ and $C_p = C_p(0)$ are singular in general. Let $X_p(\lambda)$ and X_p be their normalisations, $J_p(\lambda)$ and J_p the Jacobians of $X_p(\lambda)$ and X_p, respectively. The desingularisation is computed in detail by Archinard. As r, s and t are prime to p, the branch points have exactly one preimage in the desingularisation, by [Arc03b, Remark 3]. This makes the maps $X(\lambda)^{\mathrm{an}} \to C_p(\lambda)^{\mathrm{an}}$ and $X_p^{\mathrm{an}} \to C_p^{\mathrm{an}}$ homeomorphisms.

Remark 19.7. By replacing y by $\pm\lambda^{-t/p}y$ we get the equation for $C_p(\lambda)$ in the form

$$y^p = x^r(x-1)^s(x-\lambda^{-1})^t$$

considered in [Arc03b].

Lemma 19.8. *The genus of $X_p(\lambda)$ and X_p is $p-1$ and $(p-1)/2$, respectively.*

Proof Apply [Arc03b, Theorem 4.1] to our case.   □

The group μ_p of roots of unity operates on $C_p(\lambda)$ and C_p via $\sigma(x,y) = (x, \zeta^{-1}y)$ for $\zeta \in \mu_p$. This operation induces an operation of μ_p on $X_p(\lambda)$ and defines an embedding $\mathbb{Q}(\mu_p) \to \mathrm{End}_{\mathbb{Q}}(J_p(\lambda))$.

Lemma 19.9. *The abelian variety $J_p(\lambda)$ has at most two isotypical components. If it has two components, both have complex multiplication by $\mathbb{Q}(\mu_p)$. The abelian variety J_p has complex multiplication.*

Proof For J_p, we have $2\dim(J_p) = [\mathbb{Q}(\mu_p) : \mathbb{Q}]$, making it CM.

For $J_P(\lambda)$, the CM-field $\mathbb{Q}(\mu_p)$ operates on each isotypical component. The dimension of such a component is at least $(p-1)/2$. This implies that there is either a single isotypical component or there are two. If there are two, both factors have dimension $(p-1)/2$, making them CM.   □

Remark 19.10. The curve C_p has a cover by the Fermat curve with affine equation

$$x_1^p + x_2^p = 1$$

via $(x_1, x_2) \mapsto (x_1^p, x_1^r x_2^s)$; see Gross [Gro20]. This makes J_p (up to isogeny) a direct factor of the Jacobian of the Fermat curve. The latter has been studied intensely by Gross and Rohrlich in [Gro79].

19.2.3 Differentials on $C_p(\lambda)$

As the genus is $p-1$, the space $V = \Omega(X_p(\lambda))$ of global differential forms has dimension $p-1$ and $H^1_{\mathrm{dR}}(X_p(\lambda))$ has dimension $2(p-1)$. Recall that the latter has a description in terms of differentials of the second kind; see Lemma 14.7. For X_p, the numbers have to be divided by 2.

Lemma 19.11. *The following properties of the differential forms hold.*

1. *Suppose that* $(p, r+s+t) = 1$. *For* $1 \le n \le p-1$ *and* $0 \le u, v, w$, *the differential forms on* $X_p(\lambda)$,

$$\omega_n^{u,v,w} = \frac{x^u(1-x)^v(1-\lambda x)^w}{y^n}dx,$$

are of the second kind. They are holomorphic if and only if

$$u \ge \left[\frac{rn}{p}\right], \quad v \ge \left[\frac{sn}{p}\right], \quad w \ge \left[\frac{tn}{p}\right],$$

$$u + v + w \le \frac{n(r+s+t)-1}{p} - 1.$$

2. *Assume that* $(p, r+s) = 1$. *For* $1 \le n \le p-1$ *and* $u, v \ge 0$ *the differential forms on* X_p,

$$\omega_n^{u,v} = \frac{x^u(1-x)^v}{y^n}dx,$$

are of the second kind. They are holomorphic if and only if

$$u \ge \left[\frac{rn}{p}\right], \quad v \ge \left[\frac{sn}{p}\right],$$

$$u + v \le \frac{n(r+s)-1}{p} - 1.$$

Proof   This is in [Arc03b, Remark 12]; see also [AO18, Lemma 2.2]. In these references the lower bound is given in the form $(rn+1)/p - 1$ etc. As u is an integer, we can replace the bound by $\lceil (rn+1)/p \rceil - 1$. Let $rn = kp + e$ with $0 \le e < p$. This gives

$$\left\lceil \frac{rn+1}{p} \right\rceil - 1 = k + \left\lceil \frac{e+1}{p} \right\rceil - 1.$$

As $1 \le e+1 \le p$, we have $\lceil (e+1)/p \rceil = 1$. We may write the bound in the shape that we used. □

The forms $\omega_n^{u,v,w}$ and $\omega_n^{u,v}$ are ζ^n-eigenvalues for the operation of σ.

Corollary 19.12 (Asakura and Otsubo [AO18, Proposition 2.3]). *If* $s + t = p$, *then on* $X_p(\lambda)$

$$\omega_n := \omega_n^{u,v,w},$$

with

$$u = \left[\frac{nr}{p}\right], \quad v = \left[\frac{ns}{p}\right], \quad w = \left[\frac{nt}{p}\right],$$

is of the first kind, and

$$\eta_n := \omega_n^{u,v+1,w}$$

is of the second kind, but not of first kind. The tuple $(\omega_n | n = 1, \ldots, p - 1)$ *is a basis of* $\Omega(X_p(\lambda))$, *and* $(\omega_n, \eta_n \mid n = 1, \ldots, p - 1)$ *is a basis for* $H^1_{\mathrm{dR}}(X_p(\lambda))$.

Proof We show that these choices are the only ones giving a holomorphic form. Division with remainder gives $nr = kp + e$, $ns = lp + f$ and $nt = mp + g$. It is required that $u \geq m$, $v \geq l$, $w \geq m$. In conclusion,

$$u + v + w \geq k + l + m.$$

Furthermore it is also required that

$$u + v + w \leq \frac{n(r + s + t) - 1}{p} - 1.$$

The condition that $r + s = p$ forces p to divide $f + g$ and leads to $f + g = p$. This implies that

$$u + v + w \leq k + l + m + 1 + \frac{(e - 1)}{p} - 1$$

and since $u + v + w$ is an integer we conclude that

$$u + v + w \leq k + l + m.$$

It follows that $u + v + w = k + l + m$, $u = k$, $v = l$, $w = m$ and this is what was stated. $\qquad \square$

As the forms are eigenforms for the μ_p-operation, this is even an eigenbasis. The forms ω_n, η_n are a basis for the ζ^n-eigenspace.

Corollary 19.13. *For* $1 \leq n \leq p - 1$, *the differentials*

$$\omega_n := \frac{x^u (1 - x)^v}{y^n} dx,$$

with

$$u = \left[\frac{nr}{p}\right], \quad v = \left[\frac{ns + 1}{p}\right],$$

are a basis for $H^1_{\mathrm{dR}}(X_p)$. *The expression*

$$\left\langle \frac{nr}{p} \right\rangle + \left\langle \frac{ns}{p} \right\rangle - \left\langle \frac{n(r + s)}{p} \right\rangle, \tag{19.7}$$

where $\langle x \rangle$ denotes the fractional part of a rational number x, takes the value 1 *if ω_n is holomorphic and the value* 0 *if it is not.*

Precisely one of the forms ω_n and ω_{p-n} is holomorphic.

Proof Both forms are linearly independent because they have different eigen-values for the μ_p-operation. They span a subspace of $H^1_{\mathrm{dR}}(X_p)$ of dimension $p - 1$, hence they even generate it.

The criterion for being holomorphic is the computation of the dimension of ζ^n-eigenspace of $\Omega(X_p)$ in [Arc03b, Theorem 6.7]. We check it by hand.

Euclidean division gives $nr = kp + e$, $ns = lp + f$, and then $n(r + s) = (k + l)p + e + f$. We have $0 < e + f < 2p$ and excluded $p \mid e + f$, hence $\langle n(r + s)/p \rangle$ takes the value $(e + f)/p$ if $e + f < p$ and the value $(e + f - p)/p$ if $e + f > p$. This gives

$$\left\langle \frac{nr}{p} \right\rangle + \left\langle \frac{ns}{p} \right\rangle - \left\langle \frac{n(r + s)}{p} \right\rangle = \begin{cases} 0 & e + f < p, \\ 1 & e + f > p. \end{cases}$$

On the other hand, the condition in Lemma 19.11 contains the upper bound

$$\frac{n(r + s) - 1}{p} - 1 = \left[\frac{rn}{p} \right] + \left[\frac{rs}{p} \right] + \frac{e + f - p - 1}{p}.$$

We have $e + f - p - 1 \geq 0$ if and only if $e + f > p$. This makes $\omega_n^{u,v}$ holomorphic if and only if (19.7) takes the value 1.

The last statement is [Arc03b, Theorem 6.8]; see also [Arc03b, Remark 13]. It is also obvious from the above computation: the remainder of $(p - n)r$ is $p - e$ and the remainder of $(p - n)s$ is $p - f$.   □

A CM-pair (J, ι) consisting of an abelian variety J and an embedding $\iota \colon F \to \mathrm{End}_{\mathbb{Q}}(J)$ of a CM-field is uniquely determined up to isogeny by the pair (J, Φ), where Φ is the set of eigenvalues for the operation of F on $\Omega(J)$, the CM-type. As a by-product, the corollary describes the CM-type Φ_p of J_p. In detail: we introduce

$$H = \left\{ n \in (\mathbb{Z}/p\mathbb{Z})^* \;\middle|\; \left\langle \frac{nr}{p} \right\rangle + \left\langle \frac{ns}{p} \right\rangle - \left\langle \frac{n(r + s)}{p} \right\rangle = 1 \right\}$$

and

$$W = \{ a \in (\mathbb{Z}/p\mathbb{Z})^* \mid aH = H \}.$$

For a given $n \in H$, the condition in the corollary is satisfied. This means that ω_n is holomorphic and that ζ^n appears as an eigenvalue in the operation of μ_p on $\Omega(J_p)$. We can identify H and Φ_p. The group W is its stabiliser under the identification of the Galois group $\mathrm{Gal}(\mathbb{Q}(\mu_p)/\mathbb{Q})$ with $(\mathbb{Z}/p\mathbb{Z})^*$.

Remark 19.14. Following Gross and Rohrlich in [GR78], we may introduce a fake variable t with $r + s + t = p$ (unrelated to the t appearing $C_p(\lambda)$). Then

$$H = \left\{ n \in (\mathbb{Z}/p\mathbb{Z})^* \,\middle|\, \left\langle \frac{nr}{p} \right\rangle + \left\langle \frac{ns}{p} \right\rangle - \left\langle \frac{-nt}{p} \right\rangle = 1 \right\}$$

$$= \left\{ n \in (\mathbb{Z}/p\mathbb{Z})^* \,\middle|\, \left\langle \frac{nr}{p} \right\rangle + \left\langle \frac{ns}{p} \right\rangle + \left\langle \frac{nt}{p} \right\rangle = 2 \right\}$$

because $\langle -x \rangle = 1 - \langle x \rangle$. This is the complement in $(\mathbb{Z}/p\mathbb{Z})^*$ of the set of

$$H_{r,s,t} = \left\{ n \in (\mathbb{Z}/p\mathbb{Z})^* \,\middle|\, \left\langle \frac{nr}{p} \right\rangle + \left\langle \frac{ns}{p} \right\rangle + \left\langle \frac{nt}{p} \right\rangle = 1 \right\}$$

appearing in [GR78] because the expression takes the values 1 and 2. The normalisation of the operation of μ_p on C_p in [GR78] is complex conjugate to ours, hence they describe the CM-type by $H_{r,s,t}$ rather than by our H. Note that an $a \in (\mathbb{Z}/p\mathbb{Z})^*$ which stabilises H also stabilises $H_{r,s,t} = (\mathbb{Z}/p\mathbb{Z})^* \setminus H$, and conversely. So we actually have

$$W = W_{r,s,t},$$

as defined in [GR78, Lemma 1.6].

Lemma 19.15 (Gross and Rohrlich [GR78, Lemma 1.6]). *The group W is trivial unless $r^3 \equiv s^3 \equiv (-r - s)^3 \bmod p$, in which case it is the group of cube roots of unity (modulo p).*

Corollary 19.16. *The CM-abelian variety J_p is simple if and only if the group W is trivial. In particular, this is the case if $p \not\equiv 1 \bmod 3$.*

Proof We identify H with the CM-type, where $n \in (\mathbb{Z}/p\mathbb{Z})^*$ stands for the ζ^n-eigenspace of the operation of $\zeta \in \mu_p$ on $\Omega(J_p)$. The operation of the Galois group $\mathrm{Gal}(\mathbb{Q}(\mu_p)/\mathbb{Q}) = (\mathbb{Z}/p\mathbb{Z})^*$ on the eigenvalues is identified with the left multiplication of $(\mathbb{Z}/p\mathbb{Z})^*$ on itself. The condition in the corollary means that the stabiliser of H is trivial, i.e. the CM-type is primitive. Being primitive is equivalent to the abelian variety being simple.

If J_p is not simple, then Lemma 19.15 implies that 3 divides $p - 1$. $\quad\square$

Example 19.17. Let $p = 11$, $r = s = 2$. Then $H = \{3, 4, 5, 9, 10\}$ and J_p is simple. For $p = 7$, $r = 2$, $s = 4$, we have $H = \{3, 5, 6\}$, and J_p is not simple because $W = \{1, 2, 4\}$. For $p = 7$, $r = s = 2$, we have $H = \{2, 3, 6\}$, and J_p is again simple. These examples are compatible with the criterion of Lemma 19.15.

Remark 19.18. It can happen that $\omega_n^{u,v,w}$ is of the first kind, but $\omega_n^{u,v}$ (for the same parameters r, s, t, u, v) is not. Indeed, it must happen because the dimension of the space of differential forms of the first kind goes down.

19.2.4 Transcendence

We can now connect the periods of $X_p(\lambda)$ to our Euler integrals. Recall the differential forms $\omega(a,b,c;\lambda)$ in the complex plane; see (19.6)

Proposition 19.19. *Let p be an odd prime, $0 < r,s,t < p$ and $p \nmid r + s + t$. Choose $1 \leq n \leq p - 1$, $u,v,w \geq 0$ and introduce*

$$a = -w + \frac{nt}{p}, \quad b = u + 1 - \frac{nr}{p}, \quad c = b + 1 + v - \frac{ns}{p} = u + v + 2 - \frac{n(r+s)}{p}.$$

1. *The form $\omega_n^{u,v,w}$ is the pull-back of $\omega(a,b,c;\lambda)$ to $X_p(\lambda)$ under the projection $(x,y) \mapsto x$. The Euler integral $\Omega(a,b,c;\lambda)$ is a complete period of the second kind for $J_p(\lambda)$.*
2. *The form $\omega_n^{u,v}$ is the pull-back of $\omega(b,c-b)$ to X_p under the projection $(x,y) \mapsto x$. The Beta integral $\mathsf{B}(b,c-b)$ is a complete period of the second kind for J_p.*

Proof   The first claim is straightforward by replacing y in $\omega_n^{u,v,w}$ by

$$y = x^{r/p}(1-x)^{s/p}(1-\lambda x)^{t/p}.$$

We write ω for our meromorphic differential form. A chosen generator $\zeta \in \mu_p$ operates via σ on $C_p(\lambda)$ and (by functoriality) on $X_p(\lambda)$. Let γ be the lift of $[0,1]$ to $X_p(\lambda)$ corresponding to the choice of branch in the Euler integral. Then $(\sigma_* \gamma)^{-1}\gamma$ is a closed loop in $X_p(\lambda)^{\mathrm{an}}$. In computing the integral over the closed path we get

$$\int_\gamma \omega - \int_{\sigma_* \gamma} \omega = \int_\gamma (\omega - \sigma^* \omega) = (1 - \zeta^n) \int_\gamma \omega.$$

This makes the Euler integral a closed period of the second kind. The argument for the Beta integral is analogous. The period is of the second kind because the differential form $\omega_n^{u,v}$ is of the second kind in general.   □

By Proposition 19.19, the formula

$$\Omega(a,b,c;\lambda) = F(a,b,c;\lambda)\mathsf{B}(b,c-b)$$

of Proposition 19.6 can be regarded as a linear relation between periods of algebraic curves. The dimension computations in Chapter 15 tell us about $\overline{\mathbb{Q}}$-linear independence of period numbers.

Recall that Φ_p is the CM-type of J_p with the operation of $\mathbb{Q}(\mu_p)$ induced from the operation of μ_p on C_p. For $\alpha \in \mathrm{Gal}(\mathbb{Q}(\mu_p)/\mathbb{Q})$ we write $J_p^\alpha = (J_p, \alpha\Phi_p)$ and $\overline{J_p^\alpha} = (J_p, \overline{\alpha\Phi_p})$ (the complementary CM-type).

Theorem 19.20. *Let p be an odd prime, $0 < r, s < p$ such that p does not divide $r + s$ and put $t = p - s$. We assume that J_p is simple and take $0 \le u, v$ and a, b, c as in Proposition 19.19, $\lambda \ne 0, 1$ algebraic. If $F(a, b, c; \lambda)$ is algebraic and not zero then (up to isogeny)*

$$J_p(\lambda) \cong J_p^\alpha \times \overline{J_p^\alpha}$$

in the category of abelian varieties with μ_p-action.

Proof The assumption $t + s = p$ ensures that $p \nmid r + s + t$ is satisfied. The period $B(b, c - d)$ does not vanish, by the criterion of Summary 14.11, because the lift γ of $[0, 1]$ to X_p^{an} is not closed. As $F(a, b, c; \lambda)$ is assumed non-zero, this also makes $\Omega(a, b, c; \lambda)$ non-zero.

We consider $A = J_p \times J_p(\lambda)$. The dimension formula in Theorem 15.3 implies that there are no $\overline{\mathbb{Q}}$-linear relations between non-trivial periods of different isotypical components. This implies that J_p and $J_p(\lambda)$ share a simple factor. As J_p is simple, this means that

$$J_p(\lambda) \cong J_1 \times J_2,$$

with

$$J_p \cong J_1$$

up to isogeny. Note that we do not know if the isomorphism is compatible with the μ_p-operation.

By Lemma 19.9, both factors have CM by $\mathbb{Q}(\mu_p)$ (because either $J_p(\lambda) = J_p^2$ or it has two isotypical components) and the eigenbasis computation in Corollary 19.12 shows that J_1 and J_2 have complex conjugate CM-types Φ and $\bar{\Phi}$. As J_1 is simple, the simplicity criterion of Corollary 19.16 applies to Φ and then also to $\bar{\Phi}$. Primitive CM-types classify pairs (B, ι) for B a simple abelian variety and $\iota \colon \mathbb{Q}(\mu_p) \to \mathrm{End}_{\mathbb{Q}}(B)$. The CM-type of J_2 is realised by $(J_1, \bar{\iota})$, hence J_2 and J_1 are isogenous after all. Here $\bar{\iota}$ means complex conjugation on $\mathbb{Q}(\mu_p)$ followed by ι.

We now have $J_p(\lambda) \cong J_p^2$. Let ζ be a generator of μ_p and σ the corresponding automorphism of $J_p(\lambda)$. It operates on J_p^2 by a (2×2)-matrix with entries in $\mathbb{Q}(\mu_p) = \mathrm{End}_{\mathbb{Q}}(J_p)$. As $\sigma^p = \mathrm{id}$, it is diagonalisable and its eigenvalues are pth roots of unity. Without loss of generality, it has the shape

$$\begin{pmatrix} \zeta^\alpha & 0 \\ 0 & \zeta^{\alpha'} \end{pmatrix}.$$

The induced operation on $\Omega(J_p(\lambda))$ is diagonalisable with $p - 1$ distinct eigenvalues. This implies that $\alpha = -\alpha' \mod p$. In other words, the isomorphism $J_p(\lambda) \cong J_p^2$ can be chosen such that σ operates via ζ^α on the first factor and via

$\zeta^{-\alpha}$ on the second. This determines the CM-types of the two factors. They are complex conjugate to each other.   □

Remark 19.21. As a consequence of the theorem, the argument λ has to be *special* in the sense of Shimura varieties. This could certainly be investigated in more detail. However, the subtleties of Shimura varieties and the André–Oort conjecture are beyond the scope of our book.

Corollary 19.22. *Let p, r, s, t be as in the theorem. Let $1 \leq n \geq p - 1$ and u, v, w be as in Corollary 19.12 and a, b, c as in Proposition 19.19. If*

$$\left\langle \frac{nr}{p} \right\rangle + \left\langle \frac{ns}{p} \right\rangle - \left\langle \frac{n(r+s)}{p} \right\rangle \neq 1$$

then $F(a, b, c; \lambda)$ is zero or transcendental for all algebraic $\lambda \neq 0, 1$.

Proof In this case, the form $\omega_n^{u,v,w}$ on $X_p(\lambda)$ is of the first kind, whereas $\omega_n^{u,v}$ on X_p is not.

Assume $F = F(a, b, c; \lambda)$ is not zero and is algebraic for algebraic $\lambda \neq 0, 1$. As the CM-types of the two factors of $J_p(\lambda)$ are distinct, the form $\omega_n^{u,v,w}$ restricts to 0 on one of them, say the second. We may view it as a differential form of the first kind on J_p. As an element of $H_{dR}^1(J_p)$ it is $\overline{\mathbb{Q}}$-linearly independent of the form $\omega_n^{u,w}$, which is not of the first kind by our choice of n. By Lemma 15.6, this makes the periods of the restriction of $\omega_n^{u,v,w}$ and of $\omega_n^{u,w}$ $\overline{\mathbb{Q}}$-linearly independent.   □

Recall that we have an easy numerical criterion to check whether J_p is simple; see Lemma 19.15.

Example 19.23. We choose $p = 11, r = s = 2, r + s = 4, t = 9$. As pointed out in Example 19.17, this makes J_p simple with CM-type given by $H = \{3, 4, 5, 9, 10\}$. Conversely, ω_n is not holomorphic for $n = 1, 2, 6, 7, 8$. We have $a = \langle 9n/11 \rangle$, $b = 1 - \langle 2n/11 \rangle$, $c = 2b$

n	u	v	w	a	b	c
1	0	0	0	9/11	9/11	18/11
2	0	0	1	7/11	7/11	14/11
6	1	1	4	10/11	10/11	20/11
7	1	1	5	8/11	8/11	16/11
8	1	1	6	6/11	6/11	12/11

The corresponding values of the hypergeometric function are zero or transcendental for all algebraic $\lambda \neq 0, 1$. For $\lambda \in (0, 1)$, the Euler integral is non-zero and $F(10/11, 10/11, 20/11; \lambda)$ is a transcendental number.

PART FIVE

APPENDICES

Appendix A

Nori Motives

In this appendix a bare minimum of Nori's theory of motives is reviewed, to the extent needed in the main text. For a more complete picture, see [HMS17, Section 9.1].

Our base field will be $k \subset \mathbb{C}$ and we will work with $\mathbb{Q}$-coefficients throughout. We denote by $\mathbb{Q}$-Vect the category of finite-dimensional $\mathbb{Q}$-vector spaces and more generally by E-Mod the category of finitely generated E-left modules for a finite-dimensional $\mathbb{Q}$-algebra E.

A.1 Effective Motives and Realisations

A *diagram* D is an oriented graph. A *representation* of D is a map of oriented graphs $T \colon D \to \mathcal{A}$ into an abelian category $\mathcal{A}$. It assigns an object to every vertex and a morphism to every edge. There is an abstract construction due to Nori that attaches to every representation $T \colon D \to \mathbb{Q}$-Vect a $\mathbb{Q}$-linear abelian category. It should be thought of as the abelian category generated by D inside the category $\mathbb{Q}$-Vect. For a particular choice of the diagram and the representation we obtain the category of motives.

Definition A.1 (Huber and Müller-Stach [HMS17, Definition 9.1.1]). Let Pairs^{eff} be the diagram for which

1. the vertices are triples (X, D, i), where X is an algebraic variety over k, $D \subset X$ is a closed subvariety and $i \in \mathbb{N}_0$;
2. there are two types of edges:

   - (functoriality) for every morphism of varieties $X \to X'$ mapping a subvariety $D \subset X$ to $D' \subset X'$, an edge

     $$f^* \colon (X', D', i) \to (X, D, i);$$

- (coboundary) for every triple $X \supset Y \supset Z$, an edge

$$\partial \colon (X, Y, i) \to (Y, Z, i + 1).$$

We define the *singular realisation*

$$H_{\text{sing}} \colon \text{Pairs}^{\text{eff}} \to \mathbb{Q}\text{-Vect}$$

by mapping a vertex to the singular cohomology

$$(X, D, i) \mapsto H^i_{\text{sing}}(X^{\text{an}}, D^{\text{an}}; \mathbb{Q})$$

of the datum and edges of type f^* to pull-back on cohomology and edges of type coboundary to the coboundary map in the long exact sequence in cohomology.

Theorem/Definition A.2 (Nori [HMS17, Definition 9.1.3, Theorem 9.1.10]). *There is an abelian $\mathbb{Q}$-linear category $\mathcal{MM}^{\text{eff}}_{\text{Nori}}(k, \mathbb{Q})$, the category of effective Nori motives over k, a faithful exact functor*

$$H_{\text{sing}} \colon \mathcal{MM}^{\text{eff}}_{\text{Nori}}(k, \mathbb{Q}) \to \mathbb{Q}\text{-Vect}$$

and a representation

$$H_{\text{Nori}} \colon \text{Pairs}^{\text{eff}} \to \mathcal{MM}^{\text{eff}}_{\text{Nori}}(k, \mathbb{Q})$$

such that

$$H_{\text{sing}} \circ H_{\text{Nori}} = H_{\text{sing}}$$

is an isomorphism of functors, in particular,

$$H_{\text{sing}} \circ H_{\text{Nori}}(X, D, i) = H_{\text{sing}}(X, D, i) = H^i_{\text{sing}}(X, Y; \mathbb{Q}).$$

This triple $(\mathcal{MM}^{\text{eff}}_{\text{Nori}}(k, \mathbb{Q}), H_{\text{sing}}, H_{\text{Nori}})$ is uniquely determined by the following universal property:

For any abelian $\mathbb{Q}$-linear category $\mathcal{A}$, together with a $\mathbb{Q}$-linear faithful exact functor $f \colon \mathcal{A} \to \mathbb{Q}\text{-Vect}$ and a representation $T \colon \text{Pairs}^{\text{eff}} \to \mathcal{A}$ such that

$$f \circ T \cong H_{\text{sing}},$$

there is a $\mathbb{Q}$-linear exact functor

$$\widetilde{T} \colon \mathcal{MM}^{\text{eff}}_{\text{Nori}}(k, \mathbb{Q}) \to \mathcal{A}$$

and an isomorphism of functors $f \circ \widetilde{T} \to H^*$ which extends the isomorphism on $\text{Pairs}^{\text{eff}}$.

The universal property can be summed up in a diagram:

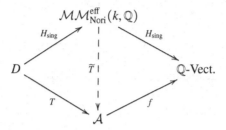

In this very precise sense, $\mathcal{MM}_{\text{Nori}}^{\text{eff}}(k,\mathbb{Q})$ is the abelian category generated by Pairs^{eff}.

We also use the notation

$$H_{\text{Nori}}^{i}(X,D) = H_{\text{Nori}}(X,D,i)$$

and call it the ith Nori motive of (X,D).

For our purposes, the most important choices for $\mathcal{A}$ are the category MHS$_k$ of mixed $\mathbb{Q}$-Hodge structures over k (see Definition 8.14), and the category $(k,\mathbb{Q})$-Vect (see Definition 7.1). Deligne constructed in [Del74] a functor $H_{\text{Hdg}}^{*} = \left(H_{\text{dR}}^{*}, H_{\text{sing}}^{*}, \phi\right)$ from the category of k-varieties to MHS$_k$. This has been extended to the diagram Pairs^{eff} (e.g. in [Hub95] or as a by-product of [Hub00, Hub04]). The situation can be summed up by the commutative diagram

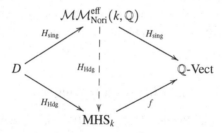

where f is the forgetful functor $(V_{\text{dR}}, V_{\text{sing}}, \phi) \mapsto V_{\text{sing}}$.

By the universal property of Nori motives, the representation H_{Hdg} extends to a functor

$$H_{\text{Hdg}}\colon \mathcal{MM}_{\text{Nori}}^{\text{eff}}(k,\mathbb{Q}) \to \text{MHS}_k$$

on Nori motives. We sum up in the following definitions.

Definitions A.3. 1. The *Hodge realisation*

$$H_{\text{Hdg}}\colon \mathcal{MM}_{\text{Nori}}^{\text{eff}}(k,\mathbb{Q}) \to \text{MHS}_k$$

is the canonical extension of the representation of Pairs^{eff} in MHS$_k$ compatible with the singular realisation.

2. The *period realisation*

$$H\colon \mathcal{MM}^{\text{eff}}_{\text{Nori}}(k,\mathbb{Q}) \to (k,\mathbb{Q})\text{-Vect}$$

is defined by forgetting the filtrations (i.e. composing with the faithful exact functor MHS$_k \to (k,\mathbb{Q})$-Vect).

3. The *de Rham realisation*

$$H_{\text{dR}}\colon \mathcal{MM}^{\text{eff}}_{\text{Nori}}(k,\mathbb{Q}) \to k\text{-Vect}$$

is defined by projecting to the k-component (i.e. composing with the faithful exact functor $(k,\mathbb{Q})$-Vect $\to k$-Vect).

Remarks A.4. 1. Every effective Nori motive M over $\overline{\mathbb{Q}}$ has a well-defined set of periods given by the periods of $H(M) \in (k,\mathbb{Q})$-Vect, i.e. the image of the period pairing

$$H_{\text{dR}}(M) \times H_{\text{sing}}(M)^{\vee} \to \mathbb{C};$$

see [HMS17, Section 11.2].

2. By the universal property of $\mathcal{MM}^{\text{eff}}_{\text{Nori}}(k,\mathbb{Q})$, every object in the category is a subquotient of a motive of the form $H^i_{\text{Nori}}(X,D)$. This will allow us to reduce questions on periods of motives to motives of the special shape.

3. In our monograph the theory of motives has been set up to be *contravariant* on the category of varieties. This follows the convention of [HMS17], but differs from Nori's original approach.

A.2 Filtration by Degree

Following Ayoub and Barbieri-Viale in [ABV15], we concentrate on the subcategories generated by motives of bounded cohomological degree or dimension.

Definition A.5. For $n \geq 0$ let $d_n\mathcal{MM}_{\text{Nori}}(k,\mathbb{Q}) \subset \mathcal{MM}^{\text{eff}}_{\text{Nori}}(k,\mathbb{Q})$ be the thick abelian subcategory (i.e. full and closed under extensions and subquotients) generated by the objects $H^i_{\text{Nori}}(X,D)$, with X a k-variety, $D \subset X$ a closed subvariety and $i \leq n$.

Remark A.6. Our definition is the contravariant analogue of [ABV15, Definition 3.1]. By [ABV15, Proposition 3.2], it suffices to deal with the case when X has dimension at most n. The period version of the argument for $n = 1$, which is the case of our interest, is given in Proposition 12.5.

For $n = 0$, the category $d_0 \mathcal{M}\mathcal{M}_{\text{Nori}}(k, \mathbb{Q})$ is the category of Artin motives; see [ABV15, Theorem 4.3]. The following theorem discusses the case $n = 1$, which is of direct relevance for us.

Theorem A.7 (Ayoub and Barbieri-Viale [ABV15, Sections 5, 6]). *The following hold.*

1. *The inclusion*

$$d_1 \mathcal{M}\mathcal{M}_{\text{Nori}}^{\text{eff}}(k, \mathbb{Q}) \to \mathcal{M}\mathcal{M}_{\text{Nori}}^{\text{eff}}(k, \mathbb{Q})$$

   has a left-adjoint.
2. *There is an anti-equivalence of categories*

$$1\text{-Mot}_k \to d_1 \mathcal{M}\mathcal{M}_{\text{Nori}}^{\text{eff}}(k, \mathbb{Q}).$$

3. *The abelian category $d_1 \mathcal{M}\mathcal{M}_{\text{Nori}}^{\text{eff}}(k, \mathbb{Q})$ can be described as the diagram category in the sense of Nori (see [HMS17, Theorem 7.1.13]), defined by the diagram with vertices (C, D), where C is a smooth affine curve and D a collection of points on C, and edges given by morphisms of pairs together with singular cohomology as a representation.*

A.3 Non-effective Motives

The full category of motives is constructed from the category of effective motives by inverting the *Lefschetz motive*

$$\mathbb{Q}(-1) = H_{\text{Nori}}^2(\mathbb{P}^1) = H_{\text{Nori}}^*(\mathbb{G}_m, \{1\}).$$

We explain the construction.

Remark A.8. Nori constructs a tensor structure on the category $\mathcal{M}\mathcal{M}_{\text{Nori}}^{\text{eff}}(k, \mathbb{Q})$. However, the resulting tensor category is not rigid. This defect can be resolved by passing to $\mathcal{M}\mathcal{M}_{\text{Nori}}(k, \mathbb{Q})$, which turns out to be rigid. We do not need the tensor structure, so we do not go into details; instead see [HMS17, Section 9.3].

The map of diagrams

$$\text{Pairs}^{\text{eff}} \to \text{Pairs}^{\text{eff}}$$

given by

$$(X, D, i) \mapsto (X \times \mathbb{G}_m, X \times \{1\} \amalg D \times \mathbb{G}_m, i + 1)$$

is compatible with the singular realisation because

$$H_{\text{sing}}^{i+1}(X \times \mathbb{G}_m, X \times \{1\} \amalg D \times \mathbb{G}_m; \mathbb{Q})$$
$$\cong H_{\text{sing}}^i(X, D; \mathbb{Q}) \otimes H^1(\mathbb{G}_m, \{1\}; \mathbb{Q}) \cong H_{\text{sing}}^i(X, D; \mathbb{Q}).$$

Note that this stupid version of the Künneth formula actually holds because the whole cohomology of $(\mathbb{G}_m, \{1\})$ is concentrated in degree 1. The datum $(\mathbb{G}_m, \{1, \}, 1)$ is what Nori calls a *good pair*.

By the universal property of the category of effective Nori motives, this induces a faithful exact functor

$$(-1)\colon \mathcal{MM}^{\mathrm{eff}}_{\mathrm{Nori}}(k, \mathbb{Q}) \to \mathcal{MM}^{\mathrm{eff}}_{\mathrm{Nori}}(k, \mathbb{Q}),$$

the *Tate twist*.

Definition A.9. The category of *Nori motives* $\mathcal{MM}_{\mathrm{Nori}}(k, \mathbb{Q})$ over k with coefficients in $\mathbb{Q}$ is defined as the localisation of $\mathcal{MM}^{\mathrm{eff}}_{\mathrm{Nori}}(k, \mathbb{Q})$ with respect to the twist functor (-1): objects in $\mathcal{MM}_{\mathrm{Nori}}(k, \mathbb{Q})$ are of the form $M(i)$ for $M \in \mathcal{MM}^{\mathrm{eff}}_{\mathrm{Nori}}(k, \mathbb{Q})$ and $i \in \mathbb{Z}$; morphisms are of the form

$$\mathrm{Hom}_{\mathcal{MM}_{\mathrm{Nori}}(k,\mathbb{Q})}(M(i), N(j)) = \lim_{n \to \infty} \mathrm{Hom}_{\mathcal{MM}^{\mathrm{eff}}_{\mathrm{Nori}}(k,\mathbb{Q})}(M(i+n), N(j+n)).$$

The functor (-1) induces an equivalence of categories

$$(-1)\colon \mathcal{MM}_{\mathrm{Nori}}(k, \mathbb{Q}) \to \mathcal{MM}_{\mathrm{Nori}}(k, \mathbb{Q}).$$

See [HMS17, Section 8.2] for a construction of the same localisation in terms of diagrams. The natural functor

$$\mathcal{MM}^{\mathrm{eff}}_{\mathrm{Nori}}(k, \mathbb{Q}) \to \mathcal{MM}_{\mathrm{Nori}}(k, \mathbb{Q})$$

is faithful because both categories are equipped with forgetful functors into the category of $\mathbb{Q}$-vector spaces.. However, we do not know if it is full.

Remark A.10. It is also an open question whether the inclusion

$$d_1 \mathcal{MM}^{\mathrm{eff}}_{\mathrm{Nori}}(k, \mathbb{Q}) \hookrightarrow \mathcal{MM}_{\mathrm{Nori}}(k, \mathbb{Q})$$

into the category of all motives is full. We give a positive answer for $k = \overline{\mathbb{Q}}$ in Theorem 13.5.

A.4 The Period Conjecture

In this section we restrict the discussion to $k = \overline{\mathbb{Q}}$.

The formalism of Chapter 7 can be applied to the diagram $\mathrm{Pairs}^{\mathrm{eff}}$ of Definition A.1 or the additive category $\mathcal{MM}^{\mathrm{eff}}_{\mathrm{Nori}}(\overline{\mathbb{Q}}, \mathbb{Q})$ and the representation/functor H with values in $(\overline{\mathbb{Q}}, \mathbb{Q})$-Vect. We make this explicit.

The periods of $\mathrm{Pairs}^{\mathrm{eff}}$ are the period numbers

$$\mathcal{P}(\mathrm{Pairs}^{\mathrm{eff}}) = \bigcup_{i=0}^{\infty} \mathcal{P}^i =: \mathcal{P}^{\mathrm{eff}}$$

of Definition 12.2. On the other hand, we have

$$\mathcal{P}(\text{Pairs}^{\text{eff}}) = \mathcal{P}(\mathcal{MM}_{\text{Nori}}^{\text{eff}}(\overline{\mathbb{Q}}, \mathbb{Q}))$$

because any object of $\mathcal{MM}_{\text{Nori}}^{\text{eff}}(\overline{\mathbb{Q}}, \mathbb{Q})$ is a subquotient of an object of the form $H_{\text{Nori}}^i(X, D)$; see Remark A.4 (2). In Definition 7.6, we also introduced the notion of a vector space of formal periods attached to an additive category or a diagram. The notion aims at a description of actual period spaces in terms of generators and relations. A priori, passing from the diagram $\text{Pairs}^{\text{eff}}$ to the abelian category $\mathcal{MM}_{\text{Nori}}^{\text{eff}}(\overline{\mathbb{Q}}, \mathbb{Q})$ might introduce new generators and more relations. However, this is not the case. We also have

$$\widetilde{\mathcal{P}}(\text{Pairs}^{\text{eff}}) = \widetilde{\mathcal{P}}(\mathcal{MM}_{\text{Nori}}^{\text{eff}}(\overline{\mathbb{Q}}, \mathbb{Q}));$$

this holds for any representation of a diagram and its diagram category. The fact is implicit in [HMS17]; see [Hub20, Theorem 3.7] for full details. Conjecture 13.1 is the Period Conjecture for $\text{Pairs}^{\text{eff}}$ in the sense of Definition 7.15, i.e. the question about injectivity of the map

$$\widetilde{\mathcal{P}}(\text{Pairs}^{\text{eff}}) \to \mathcal{P}^{\text{eff}}.$$

Up to a level of minor changes, the conjecture was formulated by Kontsevich in [Kon99]. We refer the reader to [HMS17, Remark 13.1.8] for a detailed discussion. Injectivity is equivalent to the Period Conjecture for effective Nori motives over $\overline{\mathbb{Q}}$, i.e. the injectivity of

$$\widetilde{\mathcal{P}}(\mathcal{MM}_{\text{Nori}}^{\text{eff}}(\overline{\mathbb{Q}}, \mathbb{Q})) \to \mathcal{P}^{\text{eff}}.$$

By Corollary 7.19, the Period Conjecture implies fullness of $\mathcal{MM}_{\text{Nori}}^{\text{eff}}(\overline{\mathbb{Q}}, \mathbb{Q})$. By Lemma 7.20, the Period Conjecture for $\mathcal{MM}_{\text{Nori}}^{\text{eff}}(\overline{\mathbb{Q}}, \mathbb{Q})$ is equivalent to the Period Conjecture for $\langle M \rangle$ for all $M \in \mathcal{MM}_{\text{Nori}}^{\text{eff}}(\overline{\mathbb{Q}}, \mathbb{Q})$. Here, as in Definition 7.8 we denote by $\langle M \rangle \subset \mathcal{MM}_{\text{Nori}}^{\text{eff}}(\overline{\mathbb{Q}}, \mathbb{Q})$ the smallest full subcategory containing X that is closed under subquotients. Moreover, the Period Conjecture for M can be reformulated as asking for

$$\dim_{\overline{\mathbb{Q}}} \mathcal{P}\langle M \rangle = \dim_{\mathbb{Q}} E(M)$$

with

$$E(M) := \text{End}(H_{\text{sing}}|_{\langle M \rangle}),$$

as in Definition 7.23. Unconditionally, we get the estimate

$$\dim_{\overline{\mathbb{Q}}} \mathcal{P}\langle M \rangle \leq \dim_{\mathbb{Q}} E(M).$$

In Section A.2 we have introduced the filtration of $\mathcal{MM}_{\text{Nori}}^{\text{eff}}(\overline{\mathbb{Q}}, \mathbb{Q})$ by degree.

Corollary A.11. *The Period Conjecture for all motives is equivalent to the Period Conjecture for* $d_n \mathcal{MM}_{\text{Nori}}^{\text{eff}}(\overline{\mathbb{Q}}, \mathbb{Q})$ *for all* $n \geq 0$.

Proof If $M \in d_n \mathcal{MM}_{\text{Nori}}^{\text{eff}}(\overline{\mathbb{Q}}, \mathbb{Q})$, then we even have $\langle M \rangle \subset \mathcal{MM}_{\text{Nori}}^{\text{eff}}(\overline{\mathbb{Q}}, \mathbb{Q})$. Moreover, every object of $\mathcal{MM}_{\text{Nori}}^{\text{eff}}(\overline{\mathbb{Q}}, \mathbb{Q})$ is contained in some $d_n \mathcal{MM}_{\text{Nori}}^{\text{eff}}(\overline{\mathbb{Q}}, \mathbb{Q})$. If the Period Conjecture holds for $\mathcal{MM}_{\text{Nori}}^{\text{eff}}(\overline{\mathbb{Q}}, \mathbb{Q})$, then it holds for all $\langle M \rangle$ and hence for all $d_n \mathcal{MM}_{\text{Nori}}^{\text{eff}}(\overline{\mathbb{Q}}, \mathbb{Q})$, and conversely. □

One of the main results of the present monograph is the validity of the Period Conjecture for $d_1 \mathcal{MM}_{\text{Nori}}^{\text{eff}}(\overline{\mathbb{Q}}, \mathbb{Q})$; see Theorem 13.3. This implies the dimension formula for all 1-motives; see Corollary 9.12.

The Period Conjecture can also be formulated for non-effective Nori motives; see Section A.3. We have

$$\mathcal{P}(\mathcal{MM}_{\text{Nori}}(\overline{\mathbb{Q}}, \mathbb{Q})) = \mathcal{P}^{\text{eff}}[1/2\pi i]$$

because $2\pi i$ is the period of $\mathbb{Q}(-1)$. As we do not know if $\mathcal{MM}_{\text{Nori}}^{\text{eff}}(\overline{\mathbb{Q}}, \mathbb{Q}) \rightarrow \mathcal{MM}_{\text{Nori}}(\overline{\mathbb{Q}}, \mathbb{Q})$ is full, it is also an open question whether $\widetilde{\mathcal{P}}(\mathcal{MM}_{\text{Nori}}^{\text{eff}}(\overline{\mathbb{Q}}, \mathbb{Q})) \rightarrow \widetilde{\mathcal{P}}(\mathcal{MM}_{\text{Nori}}(\overline{\mathbb{Q}}, \mathbb{Q}))$ is injective. By Proposition 7.17, this injectivity is a consequence of the Period Conjecture for $\mathcal{MM}_{\text{Nori}}(\overline{\mathbb{Q}}, \mathbb{Q})$. We deduce it for the category $d_1 \mathcal{MM}_{\text{Nori}}(\overline{\mathbb{Q}}, \mathbb{Q})$ in Theorem 13.5.

Appendix B

Voevodsky Motives

An alternative approach to the theory of motives starts out with a triangulated category of motives, thought of as the bounded derived category of motives. We use Voevodksy's approach. His category and the results known about its relation to 1-motives are used only in the proof of Theorem 13.3 (2). We refer the reader to Voevodsky's survey in [Voe00] for explicit constructions and proofs.

B.1 Geometric Motives and the Singular Realisation

Let k be a field of characteristic 0. We work with $\mathbb{Q}$-coefficients throughout. This assumption makes most of the subtleties of [Voe00], such as finite correspondences and the distinction between the Nisnevich and the étale topology, unnecessary. There are alternative constructions by different authors which yield the same result. For example, Ayoub's category $\mathrm{DA}_c^{\mathrm{eff}}(k, \mathbb{Q})$ built with the étale topology and without transfers (see [Ayo14b]), is equivalent to Voevodksy's $\mathrm{DM}_{\mathrm{gm}}^{\mathrm{eff}}(k, \mathbb{Q})$ built with the Nisnevich topology and correspondences.

Definition B.1 (Voevodsky [Voe00, Definition 2.1.1]). We denote by $\mathrm{DM}_{\mathrm{gm}}^{\mathrm{eff}}(k, \mathbb{Q})$ the triangulated category of *effective geometric motives* and by

$$M \colon \mathrm{Var}_k \to \mathrm{DM}_{\mathrm{gm}}^{\mathrm{eff}}(k, \mathbb{Q})$$

the covariant functor which attaches to an algebraic k-variety its *(Voevodsky) motive*.

Singular cohomology extends to a contravariant functor

$$H_{\mathrm{sing}} \colon \mathrm{DM}_{\mathrm{gm}}^{\mathrm{eff}}(k, \mathbb{Q}) \to \mathbb{Q}\text{-Vect}$$

such that

$$H_{\text{sing}}(M(X)[i]) = H^i_{\text{sing}}(X, \mathbb{Q}).$$

This is a consequence and easy special case of [Hub00, Hub04].

Remark B.2. From the point of view of Voevodsky motives as well as from the point of view of 1-motives, it can be argued that it would be more natural to work with singular *homology* instead. Indeed, this is the point of view taken by Ayoub in his construction of the Betti realisation not only of motives over a field, but over general base; see [Ayo10]. On the other hand, de Rham cohomology is a lot more natural than de Rham homology. We stick to the conventions set up in [HMS17].

Let Pairs^{eff} be the diagram considered in Definition A.1 in order to define Nori motives.

Lemma B.3. *The functor M extends to a representation*

$$M \colon \text{Pairs}^{\text{eff}} \to \text{DM}_{\text{gm}}(k, \mathbb{Q})$$

such that $H_{\text{sing}} \circ M$ agrees with the singular realisation of Section A.1.

Proof Let (X, Y, i) be a vertex of Pairs^{eff}, i.e. X is an algebraic variety, Y a closed subvariety and $i \geq 0$. We define

$$M(X, Y, i) = M(Y \to X)[i],$$

where $M(Y \to X)$ is the object of $\text{DM}_{\text{gm}}^{\text{eff}}(k, \mathbb{Q})$ corresponding to the bounded complex $[Y \to X]$ in the additive category SmCor of finite correspondences; see [Voe00, Section 2.1]. Its singular realisation is $H^i_{\text{sing}}(X, Y; \mathbb{Q})$.

A morphism $f \colon X \to X'$ mapping Y to Y' induces a natural morphism $f^* \colon M(X', Y', i) \to M(X, Y, i)$. A triple $Z \subset Y \subset X$ induces

$$\partial \colon M(X, Y, i) \to M(Y, Z, i + 1).$$

Both are mapped to the correct map on singular cohomology. □

B.2 Filtration by Dimension

Definition B.4 (Voevodsky [Voe00, Section 3.4]). For $n \geq 0$ let $d_n \text{DM}_{\text{gm}}^{\text{eff}}(k, \mathbb{Q})$ be the full thick subcategory generated by the motives of the form $M(X)$ for a smooth variety X of dimension at most n.

The case $n = 0$ is easy. Voevodsky showed that $d_0 \text{DM}_{\text{gm}}^{\text{eff}}(k, \mathbb{Q})$ is equivalent to the bounded derived category of finite-dimensional continuous representations of $\text{Gal}(\bar{k}/k)$. If k is algebraically closed, this is simply the bounded derived category of $\mathbb{Q}$-Vect.

Theorem B.5 (Orgogozo [Org04], Barbieri-Viale–Kahn [BVK16]). *There is a natural equivalence of triangulated categories*

$$D^b(1\text{-Mot}_k) \to d_1 \text{DM}_{\text{gm}}(k, \mathbb{Q})$$

from the derived category $D^b(1\text{-Mot}_k)$ of the abelian category of iso-1-motives to $d_1 \text{DM}_{\text{gm}}(k, \mathbb{Q})$. The inclusion $d_1 \text{DM}_{\text{gm}}(k, \mathbb{Q}) \to \text{DM}_{\text{gm}}^{\text{eff}}(k, \mathbb{Q})$ has a left adjoint which is a section.

For $n \geq 2$, the subcategories $d_n \text{DM}_{\text{gm}}^{\text{eff}}(k, \mathbb{Q})$ remain mysterious.

B.3 Relation to Nori Motives

We have now seen two approaches to a theory of motives. They are related.

Theorem B.6 (Nori, Harrer [Har16]). *There is a triangulated functor*

$$\text{DM}_{\text{gm}}^{\text{eff}}(k, \mathbb{Q}) \to D^b(\mathcal{M}\mathcal{M}_{\text{Nori}}^{\text{eff}}(k, \mathbb{Q}))$$

between triangulated categories compatible with the singular realisation into the derived category of $\mathbb{Q}$-vector spaces.

Proof The existence of the functor is due to Harrer, based on a construction of Nori. $\square$

Proposition B.7. *The functor of Theorem B.6 maps $d_n \text{DM}_{\text{gm}}^{\text{eff}}(k, \mathbb{Q})$ to the full subcategory of $D^b(\mathcal{M}\mathcal{M}_{\text{Nori}}^{\text{eff}}(k, \mathbb{Q}))$ consisting of objects with cohomology in $d_n \mathcal{M}\mathcal{M}_{\text{Nori}}^{\text{eff}}(k, \mathbb{Q})$.*

Proof As $d_n \mathcal{M}\mathcal{M}_{\text{Nori}}^{\text{eff}}(k, \mathbb{Q})$ is closed under subquotients and extensions, it suffices to check the claim for a system of generators for $d_n \text{DM}_{\text{gm}}^{\text{eff}}(k, \mathbb{Q})$ as a triangulated category. We first do the case $n = 1$, which is of most significance for us.

Let C be a connected smooth variety of dimension 0 or 1 and $M(C)$ be the corresponding object in $\text{DM}_{\text{gm}}(k, \mathbb{Q})$. Its image in $D^b(\mathcal{M}\mathcal{M}_{\text{Nori}}(k, \mathbb{Q}))$ has cohomology in degree at most 2. Cohomology in degree 2 occurs only if C is a smooth proper curve. In this case

$$H_{\text{Nori}}^2(C) \cong H_{\text{Nori}}^2(\mathbb{P}^1) \cong H_{\text{Nori}}^1(\mathbb{G}_m),$$

hence it is also in $d_1 \mathcal{M}\mathcal{M}_{\text{Nori}}(k, \mathbb{Q})$.

In the general case, we need to consider $M(X)$ with X a smooth variety of dimension at most n. It remains to show that $H_{\text{Nori}}^i(X)$ is in $d_n \mathcal{M}\mathcal{M}_{\text{Nori}}^{\text{eff}}(k, \mathbb{Q})$.

By the Mayer–Vietoris property, we may assume that X is affine. We then follow the construction of $H^i_{\mathrm{Nori}}(X)$ in [HMS17]. We choose a *good filtration*

$$X_0 \subset X_1 \subset \cdots \subset X_n = X$$

by closed subvarieties such that $H^i_{\mathrm{sing}}(X_j, X_{j-1}; \mathbb{Q})$ is concentrated in degree j. (The existence of such a filtration is guaranteed by Nori's Basic Lemma; see [HMS17, Proposition 9.2.3].) This yields the complex

$$H^0_{\mathrm{Nori}}(X_0) \to H^1_{\mathrm{Nori}}(X_1, X_0) \to \cdots \to H^n_{\mathrm{Nori}}(X_n, X_{n-1}),$$

whose cohomology is equal to $H^*_{\mathrm{Nori}}(X)$. As the complex is in $d_n \mathcal{M}\mathcal{M}^{\mathrm{eff}}_{\mathrm{Nori}}(k, \mathbb{Q})$, so is its cohomology. $\qquad\qquad\qquad\qquad\qquad\qquad\qquad\qquad\qquad\qquad$ $\square$

Appendix C

Comparison of Realisations

In this appendix, we identify Deligne's explicit realisations of 1-motives with realisations of motives of algebraic varieties.

We work over an algebraically closed field k with a fixed embedding into $\mathbb{C}$. Recall the functor

$$V \colon \text{1-Mot}_k \to \text{MHS}_k \to (\mathbb{Q}, k)\text{-Vect}.$$

It maps the iso-1-motive $M = [L \to G]$ to the triple consisting of its singular realisation $V_{\text{sing}}(M)$, its de Rham realisation $V_{\text{dR}}(M)$ and the period isomorphism. We refer the reader to Chapter 8 for their construction.

On the other hand, there is a functor

$$(X, Y) \mapsto H^1(X, Y) \in (k, \mathbb{Q})\text{-Vect}$$

for pairs of a variety X and a closed subvariety. A complete reference for its construction is [HMS17] and this is what we are going to rely on. We have already discussed it in Appendix A in the context of Nori motives. Alternatively, we can also construct it from the theory of Voevodsky motives; see Appendix B: every morphism $Y \to X$ of smooth k-varieties gives rise to an object $M(Y \to X) \in \text{DM}_{\text{gm}}(k)$. Let $H^0 \colon \text{DM}_{\text{gm}}(k) \to (k, \mathbb{Q})\text{-Vect}$ be the standard cohomological functor of [Hub00, Hub04] and let $H^1 = H^0 \circ [1]$. By composition, we define

$$H^1(Y \to X) \in (k, \mathbb{Q})\text{-Vect}.$$

Its de Rham component is defined explicitly as H^1 of

$$R\Gamma_{\text{dR}}(X, Y) := \text{cone}(R\Gamma_{\text{dR}}(Y) \to R\Gamma_{\text{dR}}(X))[-1]$$

and its singular component as H^1 of

$$R\Gamma_{\text{sing}}(X, Y) := \text{cone}(R\Gamma_{\text{sing}}(Y) \to R\Gamma_{\text{sing}}(X))[-1],$$

where $R\Gamma_{dR}$ and $R\Gamma_{sing}$ are functorial complexes computing de Rham and singular cohomology respectively; see [HMS17, Sections 3.3.3, 5.5]. They are connected by a period isomorphism; see [HMS17, Corollary 5.52]. In the special case $Y \subset X$, we get back relative cohomology.

C.1 The de Rham Realisation Revisited

In Chapter 3 we gave an explicit definition of relative algebraic de Rham cohomology in the smooth case. We need to relate it to the one in our main reference [HMS17].

Lemma C.1. *Let X be a smooth variety and $\mathfrak{U}$ a finite open affine cover of X. Then there is a natural isomorphism of complexes in the derived category*

$$R\widetilde{\Gamma}_{dR}(X, \mathfrak{U}) \to R\Gamma_{dR}(X),$$

functorial for affine maps. If $Y \subset X$ is a smooth closed subvariety, then

$$R\widetilde{\Gamma}_{dR}(X, Y, \mathfrak{U}) := \text{cone}(R\Gamma_{dR}(Y, \mathfrak{U} \cap Y) \to R\Gamma_{dR}(X, \mathfrak{U}))[-1]$$

is naturally isomorphic to $R\Gamma_{dR}(X, Y)$.

Proof The definition of $R\Gamma_{dR}$ for complexes of smooth varieties is given in [HMS17, Definition 3.3.1]. For a single smooth variety, by [HMS17, Definition 3.3.14], it is given as global sections of the Godement resolution (see [HMS17, Section 1.4.2]),

$$R\Gamma_{dR}(X) = \text{Gd}_X \Omega_X^*(X).$$

The Čech-complex

$$\text{tot}(C^*(\mathfrak{U}, \text{Gd}_X \Omega_X^*))$$

receives natural quasi-isomorphisms both from $R\widetilde{\Gamma}(X, \mathfrak{U})$ (because the natural map $\Omega_X^* \to \text{Gd}_X \Omega_X^*$ is a quasi-isomorphism of complexes of sheaves) and from $R\Gamma_{dR}(X)$ (because the cover $\mathfrak{U}$ refines X). Together they define an isomorphism in the derived category.

The construction extends to complexes $Y \to X$. □

C.2 The Comparison Result

In Definition 7.12, we introduced the external duality functor

$$\cdot^\vee : (\mathbb{Q}, k)\text{-Vect} \to (k, \mathbb{Q})\text{-Vect}$$

mapping the triple $(V_k, V_{\mathbb{Q}}, \phi)$ to $(V_{\mathbb{Q}}^\vee, V_k^\vee, \phi^\vee)$.

Proposition C.2. *Let* $M = [L \xrightarrow{f} G]$ *be a* 1-*motive and let* $e_1, \ldots, e_r$ *be a basis of* L *and put* $e_0 = 0$. *We introduce* $Z = \coprod_{i=0}^{r} \mathrm{Spec}(k) = \{P_0, \ldots, P_r\}$ *and* $\tilde{f} : Z \to G$ *given by* $\tilde{f}(P_i) = f(e_i)$. *Then*

$$V(M)^{\vee} \cong H^1(Z \xrightarrow{\tilde{f}} G).$$

The proof will take the rest of the appendix. Before going into it, we want to record a consequence. Let $M = [L \to G]$ be a 1-motive. The assignment

$$S \mapsto [L \otimes \mathbb{Q} \to G(S) \otimes_{\mathbb{Z}} \mathbb{Q}]$$

defines a complex of homotopy invariant Nisnevich sheaves with transfers on the category Sm/k of smooth k-varieties; see [AEWH15, Lemma 2.1.2], building on work of Spieß-Szamuely and Orgogozo. Hence it defines an object $\underline{L} \to \underline{G}$ in Voevodsky's category of motivic complexes with $\underline{L}$ in degree 0. In fact, it is even an object of the full subcategory $\mathrm{DM}_{\mathrm{gm}}(k, \mathbb{Q})$; see [AEWH15, Proposition 5.2.1]. We denote it $M_{\mathrm{gm}}(\underline{L} \to \underline{G})$.

Corollary C.3. *Let* $M = [L \to G]$ *be a* 1-*motive over* k. *Then*

$$V(M)^{\vee} \cong H^*(M_{\mathrm{gm}}(\underline{L} \to \underline{G})) = H^0(M_{\mathrm{gm}}(\underline{L} \to \underline{G})).$$

Proof The equality on the right is [AEWH15, Proposition 7.2.3]. In order to compare $V(M)^{\vee}$ and $H^0(M_{\mathrm{gm}}(\underline{L} \to \underline{G}))$ it suffices to give a natural isomorphism

$$H^0(M_{\mathrm{gm}}(\underline{L} \to \underline{G})) \to H^1(G, Z)$$

with Z as in Proposition C.2. The main result of [AEWH15] is to describe $\underline{G}$ as a direct summand of $M_{\mathrm{gm}}(G)$. Its cohomology in any contravariant Weil cohomology theory (e.g. de Rham or singular cohomology) agrees with $H^1(G)$. In particular,

$$H^*(M_{\mathrm{gm}}(\underline{G})) = H^1(G).$$

The projection map $M(G) \to \underline{G}$ is given by the summation map

$$M(G)(S) = \mathrm{Cor}(S, G) \to G(S).$$

On the other hand, $\underline{L} = M(Z')$ with $Z' = \{P_1, \ldots, P_r\}$. There is a natural projection $M(Z) \to M(Z')$.

Our comparison isomorphism is induced by the morphism of motivic complexes

$$[M(Z) \to M(G)] \to [M(Z') \to \underline{G}].$$

We apply the long exact cohomology sequence for the stupid filtration:

$$H^0(G) \longrightarrow H^0(Z) \longrightarrow H^1(G,Z) \longrightarrow H^1(G) \longrightarrow 0$$
$$\big\uparrow \qquad\qquad \big\uparrow \qquad\qquad \text{id}\big\uparrow$$
$$0 \longrightarrow H^0(Z') \longrightarrow H^0(M_{\mathrm{gm}}(\underline{L} \to \underline{G})) \longrightarrow H^1(G) \longrightarrow 0.$$

The vertical map on the left induces an isomorphism

$$\mathbb{Q}^r \cong H^0(Z') \to H^0(Z)/H^0(G) = \mathbb{Q}^{r+1}/\Delta(\mathbb{Q}).$$

Together with the identity on the right, this induces an isomorphism in the middle. $\qquad\qquad\qquad\qquad\qquad\qquad\qquad\qquad\qquad\qquad\quad\square$

We now start on the proof of Proposition C.2.

Lemma C.4. *Let $M = [L \to G] \in 1\text{-Mot}_k$ and $Z \to G$ be as in Proposition C.2. Then*

$$V_{\mathrm{dR}}(M)^\vee \cong H^1_{\mathrm{dR}}(Z \to G).$$

Proof We refer the reader to Chapter 8 for the construction of the algebraic group $M^\natural \to G$ and the realisations. Note that $R\Gamma_{\mathrm{dR}}(Z) = H^0(Z, \mathcal{O}_Z) \cong k^{r+1}$ because Z is of dimension zero. Hence

$$R\Gamma_{\mathrm{dR}}(G,Z)^i = \begin{cases} R\Gamma_{\mathrm{dR}}(G)^i & i \neq 1, \\ R\Gamma_{\mathrm{dR}}(G)^1 \oplus H^0(Z, \mathcal{O}_Z) & i = 1. \end{cases}$$

We claim that there is a natural map in the derived category

$$V_{\mathrm{dR}}(M)^\vee[-1] = \mathrm{coLie}(M^\natural)[-1] \to R\Gamma_{\mathrm{dR}}(G,Z).$$

It is induced by the composition

$$\mathrm{coLie}(M^\natural) \to \Omega^1(M^\natural) \to R\Gamma_{\mathrm{dR}}(M^\natural)^1 \leftarrow R\Gamma_{\mathrm{dR}}(G)^1,$$

where the maps represent the following:

- extension of an element of the cotangent space to a unique equivariant differential form (which is closed because the co-Lie bracket is trivial);
- association of a section of the de Rham complex to a section of its Godement resolution;
- functoriality of $R\Gamma_{\mathrm{dR}}$ (a quasi-isomorphism by homotopy invariance),

together with the map

$$\mathrm{coLie}(M^\natural) \xrightarrow{0} H^0(Z, \mathcal{O}_Z).$$

We compare the long exact sequence of the cone with the exact sequence for $V_{dR}(M)$:

$$0 \longrightarrow H^0_{dR}(G) \longrightarrow H^0_{dR}(Z) \longrightarrow H^1_{dR}(G,Z) \longrightarrow H^1_{dR}(G) \longrightarrow 0$$

$$0 \longrightarrow \mathrm{Hom}(L,\mathbb{G}_a) \longrightarrow \mathrm{coLie}(M^{\natural}) \longrightarrow \mathrm{coLie}(G^{\natural}) \longrightarrow 0.$$

The square on the right commutes by naturality. The dotted arrow does not exist as a natural map, but we get an induced map

$$\mathrm{Hom}(L,\mathbb{G}_a) \to H^0_{dR}(Z)/H^0_{dR}(G).$$

We make it explicit. Pick $\alpha\colon L \to \mathbb{G}_a$. It defines a differential form $\omega(\alpha)$ on the algebraic group $V = \mathbb{G}_a(\mathrm{Hom}(L,\mathbb{G}_a)^{\vee})$. In coordinates: let t_i be the coordinate on V corresponding to the basis vector e_i of L. Then

$$\omega(\alpha) = \sum_{i=1}^{r} \alpha(e_i)dt_i.$$

We have an exact sequence

$$0 \to G^{\natural} \to M^{\natural} \to V \to 0,$$

hence we can pull $\omega(\alpha)$ back to $M^{\natural}$. This defines a class in $H^1_{dR}(G,Z)$. By construction, its image in $H^1_{dR}(G)$ vanishes. This implies that the class is exact. Indeed,

$$\omega(\alpha) = d\left(\sum_{i=1}^{r} \alpha(e_i)t_i + c\right)$$

for any $c \in k$. We lift $Z \to G$ to $Z \to M^{\natural}$ by mapping P_i to the image of e_i in $M^{\natural}$. We get a class in $H^0_{dR}(Z) = H^0(Z,\mathcal{O}_Z)$ by restricting our function $\sum_i \alpha(e_i)t_i + c$ to Z:

$$P_i \mapsto \alpha(e_i) + c.$$

(Note that $\alpha(e_0) = \alpha(0) = 0$.) The equivalence class in $H^0_{dR}(Z)/H^0_{dR}(G)$ is independent of c and hence well defined.

Now that we have an explicit formula, it is obvious that the map is bijective. Hence it remains to show that $\mathrm{coLie}(G^{\natural}) \to H^1_{dR}(G)$ is an isomorphism. Both coLie and H^1_{dR} are exact functors on the category of semi-abelian varieties, hence it suffices to consider the cases $G = \mathbb{G}_m$ and $G = A$ abelian. In the first case, $\mathbb{G}_m^{\natural} = \mathbb{G}_m$ and the invariant differential dt/t is known to generate $H^1_{dR}(\mathbb{G}_m)$.

Let $G = A$ be an abelian variety. In this case, $\Omega^1(A) = \Omega^1(A)^A \cong \mathrm{coLie}(A)$. By Hodge theory, we have the short exact sequence

$$0 \to \Omega^1(A) \to H^1_{\mathrm{dR}}(A) \to H^1(A, \mathcal{O}_A) \to 0.$$

The last group also identifies with $\mathrm{Ext}^1(A, \mathbb{G}_a)$. The exact sequence is compatible with the sequence

$$0 \to \mathrm{coLie}(A) \to \mathrm{coLie}(A^\natural) \to \mathrm{Ext}^1(A, \mathbb{G}_a) \to 0.$$

Hence $\mathrm{coLie}(A^\natural) \to H^1_{\mathrm{dR}}(A)$ is an isomorphism as well.   □

Lemma C.5. *Let $M = [L \to G] \in 1\text{-Mot}_k$ and $Z \to G$ be as in Proposition C.2. Then*

$$V_{\mathrm{sing}}(M)^\vee \cong H^1_{\mathrm{sing}}(Z \to G).$$

Proof It is more natural to give the argument in terms of homology. We use the description via C^∞-chains; see Remark 3.11 or [HMS17, Definition 2.2.2]. We work with integral coefficients throughout and omit them from the notation.

Recall that, by construction, $f \colon L \to G$ has an injective lift $f^\natural \colon L \to M^\natural$. By abuse of notation, the map $P_i \mapsto f^\natural(e_i)$ is also denoted $\widetilde{f} \colon Z \to M^\natural$. By homotopy invariance, $H^{\mathrm{sing}}_1(Z \to G) \cong H^{\mathrm{sing}}_1(Z \to M^\natural)$. From now on we work with the latter.

Given an algebraic variety X/k let $S^\infty_*(X)$ be the chain complex of C^∞-chains on X^{an} with integral coefficients. For a morphism $f \colon Y \to X$, we then define $H^{\mathrm{sing}}_i(X, Y)$ as H_i of the complex

$$S^\infty_*(X, Y) := \mathrm{cone}(S^\infty_*(Y) \to S^\infty_*(X)).$$

As Z is a disjoint union of points, $S^\infty_n(Z) = \mathbb{Z}[Z]$ (linear combinations of points) for all $n \geq 0$, and the map of complexes

$$S^\infty_*(Z) \to \mathbb{Z}[Z][0]$$

is a quasi-isomorphism, where as usual $\mathbb{Z}[Z][0]$ denotes the complex concentrated in degree 0. We define a map of complexes

$$S^\infty_*(M^\natural) \to [\mathrm{Lie}(M^{\natural,\mathrm{an}}) \to M^{\natural,\mathrm{an}}]$$

as follows:

- in degree 1 a path $\gamma \colon [0,1] \to M^{\natural,\mathrm{an}}$ is mapped to $I(\gamma) \in \mathrm{Lie}(M^{\natural,\mathrm{an}})$ (see Section 5.3);
- in degree 0 a formal sum in $\mathbb{Z}[M^{\natural,\mathrm{an}}]$ is mapped to its sum in $M^{\natural,\mathrm{an}}$.

This is compatible with the differential because $\exp(I(\gamma)) = \gamma(1) - \gamma(0)$. It is easy to see that the diagram of complexes

$$
\begin{array}{ccc}
S_*^\infty(Z) & \longrightarrow & S_*^\infty(M^\natural) \\
\downarrow & & \downarrow \\
L[0] & \longrightarrow & [\mathrm{Lie}(M^{\natural,\mathrm{an}}) \to M^{\natural,\mathrm{an}}]
\end{array}
$$

commutes. Hence we get morphisms of complexes

$$
S_*^\infty(M^\natural, Z) \to \mathrm{cone}\left(L[0] \to [\mathrm{Lie}(M^{\natural,\mathrm{an}}) \to M^{\natural,\mathrm{an}}]\right)
$$
$$
= [L \oplus \mathrm{Lie}(M^{\natural,\mathrm{an}}) \to M^{\natural,\mathrm{an}}]
$$
$$
\to [\mathrm{Lie}(M^{\natural,\mathrm{an}}) \to M^{\natural,\mathrm{an}}/f^\natural(L)].
$$

By definition,

$$
T_{\mathrm{sing}}(M) = \ker(\mathrm{Lie}(M^{\natural,\mathrm{an}}) \to M^{\natural,\mathrm{an}}/f^\natural(L)),
$$

hence we have defined a natural morphism in the derived category

$$
S_*^\infty(M^\natural, Z) \to T_{\mathrm{sing}}(M)[-1].
$$

We compare the long exact sequence of the cone with the exact sequence for $V_{\mathrm{sing}}(M)$:

$$
\begin{array}{ccccccc}
0 \longrightarrow H_1^{\mathrm{sing}}(M^\natural) & \longrightarrow & H_1^{\mathrm{sing}}(Z \to M^\natural) & \longrightarrow & H_0^{\mathrm{sing}}(Z) & \longrightarrow\!\!\!\!\rightarrow & H_0^{\mathrm{sing}}(M^\natural) \\
\downarrow & & \downarrow & & \downarrow & & \\
0 \longrightarrow T_{\mathrm{sing}}(G) & \longrightarrow & T_{\mathrm{sing}}(M) & \longrightarrow & L & \longrightarrow & 0.
\end{array}
$$

We make the dotted maps explicit. The elements of $H_1^{\mathrm{sing}}(M^\natural)$ are represented by closed loops in $M^{\natural,\mathrm{an}}$. Let γ be such a closed loop. Then the endpoints of $\widetilde\gamma$ have the same image in $M^{\natural,\mathrm{an}}$, hence $\widetilde\gamma(1) - \widetilde\gamma(0) \in \ker(\mathrm{Lie}(M^{\natural,\mathrm{an}}) \to M^{\natural,\mathrm{an}}) = T_{\mathrm{sing}}(G)$. The dotted map on the left is an isomorphism.

The kernel of $H_0^{\mathrm{sing}}(Z) \to H_0^{\mathrm{sing}}(G)$ is generated by formal differences $P_i - P_0$. Choose a path γ_i from $\widetilde f(P_0) = 0$ to $\widetilde f(P_i)$ in $M^{\natural,\mathrm{an}}$. Then

$$
(-\gamma_i, P_i - P_0) \in S_1^\infty(M^\natural, Z)
$$

is in the kernel of boundary map. Hence its cohomology class is the preimage of $P_i - P_0$. Its image in $\mathrm{Lie}(M^{\natural,\mathrm{an}})$ is given by $\widetilde\gamma_i(0) - \widetilde\gamma_i(1)$ for a lift $\widetilde\gamma_i$ of γ_i. We may choose $\widetilde\gamma_i(0) = 0$, then $-\widetilde\gamma_i(1)$ is in the preimage of $\widetilde f(P_i)$. (Which preimage depends on the choice of γ_i. It is only unique up to 2-chains.) Its equivalence class modulo $T_{\text{sing}}(G)$ is nothing but the image in $M^{\natural,\text{an}}$, hence $\widetilde{f}(P_i) = f^\natural(e_i)$. The map $\ker(H_0(Z) \to H_0(M^\natural)) \to L$ is also bijective.

This completes the proof. □

Lemma C.6. *The comparison maps for de Rham and singular cohomology are compatible with the period isomorphism.*

Proof In keeping with the proof of the last lemma, we prefer to check compatibility with the period pairing:

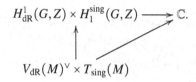

$$H^1_{\text{dR}}(G,Z) \times H^{\text{sing}}_1(G,Z) \longrightarrow \mathbb{C}.$$

$$V_{\text{dR}}(M)^\vee \times T_{\text{sing}}(M)$$

A priori, the pairing on the 1-motive level is simply evaluation of an element of $\text{coLie}(M^\natural)$ on $\sigma \in \text{Lie}(M)_{\mathbb{C}}$. However, we have already made the translation to the integration of equivariant differential forms on $M^{\natural,\text{an}}$ along paths.

We now view the same differential form as a class in de Rham cohomology and the path as a class in singular homology. The cohomological version of the period pairing is also given by integration in these special cases. □

Proof of Proposition C.2 Combine Lemmas C.4, C.5 and C.6. □

Notation

General

k	algebraically closed field with a fixed embedding $k \to \mathbb{C}$
$V^\vee$	dual vector space
$\langle \cdot, \cdot \rangle$	natural pairing between a vector space and its dual
X^{an}	analytic space attached to an algebraic variety over k
G^{an}	complex Lie group attached to an algebraic group over k
Var_k	varieties over k, i.e. reduced k-schemes of finite type
Sm_k	smooth varieties over k
$D^b(\mathcal{A})$	bounded derived category of an abelian category

Chapter 2

$\mathcal{A} \otimes \mathbb{Q}$	isogeny category of an additive category $\mathcal{A}$
$\mathbb{Z}[\mathcal{C}]$	additive hull of a category

Chapter 3

Δ_n	standard n-dimensional simplex
$S_n(X)$	space of singular n-chains
$H_n^{\mathrm{sing}}(X, \mathbb{Q})$	singular homology
$H_n^{\mathrm{sing}}(X, Y; \mathbb{Q})$	relative singular homology
$H_{\mathrm{dR}}^n(X)$	algebraic de Rham cohomology
$H_{\mathrm{dR}}^n(X, Y)$	relative algebraic de Rham cohomology
$R\widetilde{\Gamma}_{\mathrm{dR}}(X, \mathfrak{U})$	explicit complex computing de Rham cohomology

Chapter 4

$\mathcal{G}$	category of commutative connected algebraic groups over k
$\mathbb{G}_a$	additive group
$\mathbb{G}_m$	multiplicative group
$A^\vee$	dual abelian variety
$X(T)$	character group of a torus
$\mathbb{G}_m(\Xi)$	dual torus
$\mathrm{Ext}^1(A, B)$	Yoneda-extension group
$G^\natural$	universal vector extension of a semi-abelian variety
$J(Y)$	generalised Jacobian of a smooth curve

Chapter 5

$\mathrm{Lie}(G), \mathfrak{g}$	Lie algebra of an algebraic group or a complex Lie group
$\exp_G$	exponential map of commutative complex Lie group
$I(\gamma)$	path integral as inverse of the exponential map
$\log_G$	multi-valued inverse of $\exp_G$

Chapter 6

$\mathrm{Ann}(u)$	annihilator of an element

Chapter 7

(K, L)-Vect	category of pairs of vector spaces with a period isomorphism
$\mathcal{P}(V)$	set of periods of an object $V \in (K, L)$-Vect
$\mathcal{P}\langle V \rangle$	space of periods of an object $V \in (K, L)$-Vect
$\mathcal{P}(\mathcal{C})$	set of periods of a category
$\widetilde{\mathcal{P}}(\mathcal{C})$	space of formal periods of an additive category
ev	evaluation map for formal periods
$V^\vee$	external dual of $V \in (K, L)$-Vect
$\mathrm{End}(T)$	endomorphism algebra of a functor T on $\mathcal{C}$
$\mathcal{A}(\mathcal{C}, T)$	coalgebra of a functor T on $\mathcal{C}$

Chapter 8

$[L \to G]$	1-motive with lattice part L and semi-abelian part G
1-Mot_k	the category of iso-1-motives over k
$T_{\mathrm{sing}}(M)$	integral singular realisation of a 1-motive
$V_{\mathrm{sing}}(M)$	rational singular realisation of a 1-motive

$M^{\natural}$	universal vector extension of a 1-motive
1–MOT_k	category of generalised 1-motives
$V_{\mathrm{dR}}(M)$	de Rham realisation of a 1-motive
ϕ_M	period isomorphism for a 1-motive
MHS_k	the category of mixed $\mathbb{Q}$-Hodge structures over k

Chapter 9

$\mathcal{P}(M)$	set of periods of a 1-motive
$\mathcal{P}\langle M\rangle$	space of periods of a 1-motive
$\int_\sigma \omega$	period pairing for a 1-motive
$\mathrm{coLie}(G)$	dual of the Lie algebra
$V_{\mathrm{dR}}^\vee(M)$	dual of the de Rham realisation of M
$E(M)$	endomorphism algebra for the period realisation of a 1-motive
V	functor $1\text{-Mot}_{\overline{\mathbb{Q}}} \to (\mathbb{Q},\overline{\mathbb{Q}})$-Vect

Chapter 10

(GS)	Gelfond–Schneider formulation of Hilbert's seventh problem
(B)	Baker formulation of Hilbert's seventh problem

Chapter 11

$\delta(M)$	dimension of the space of periods of M

Chapter 12

$H^i(X,Y)$	object in (K,L)-Vect attached to $Y \subset X$
∂	(co)boundary map for relative (co)homology
$\mathcal{P}(X,Y,i)$	set of periods for $H^i(X,Y)$
$\mathcal{P}^i$	set of i-periods for all X,Y
$\mathbb{Z}[D]^0$	divisors of degree 0 supported on D
$J(C)$	generalised Jacobian of a smooth curve
$d_1\mathrm{DM}_{\mathrm{gm}}(k,\mathbb{Q})$	geometric motives of dimension at most 1
$d_1\mathcal{MM}_{\mathrm{Nori}}^{\mathrm{eff}}(k,\mathbb{Q})$	Nori motives of degree at most 1

Chapter 13

$\Omega^1(C)$	regular algebraic differential forms on a smooth projective curve C

Chapter 14

$l(\sigma)$	element of $\mathrm{Lie}(J(C)^{\mathrm{an}})$ attached to the chain σ
$l^\circ(\sigma)$	element of $\mathrm{Lie}(J(C^\circ)^{\mathrm{an}})$ attached to the chain σ
$[\omega]$	class of a differential form in de Rham cohomology
$[\sigma]$	class of a chain in singular homology
$\mathbb{Z}_\sigma$	subgroup of $\mathbb{Z}[D]^0$ generated by $\partial\sigma$

Chapter 15

$\delta(M)$	dimension of the period space of a 1-motive
$g(B)$	dimension of the abelian variety B
$e(B)$	dimension of $\mathrm{End}_{\mathbb{Q}}(B)$ for an abelian variety B
$\mathrm{rk}_B(L, B)$	L-rank of M with respect to the simple abelian variety B
$\mathrm{rk}_B(T, M)$	T-rank of M with respect to B
$\mathrm{rk}_{\mathbb{G}_m}(L, M)$	L-rank of a Baker motive in $\mathbb{G}_m$
X_{red}	image of the lattice $X(T)$ in $A^\vee(\overline{\mathbb{Q}})_{\mathbb{Q}}$
X_{sat}	saturation of X_{red}
G_{red}	semi-abelian variety with torus lattice X_{red}
G_{sat}	semi-abelian variety with torus lattice X_{sat}
L_{red}	image of the lattice L in $A(\overline{\mathbb{Q}})_{\mathbb{Q}}$
L_{sat}	saturation of L_{red}
M_{red}	reduced motive constructed from M
M_{sat}	saturated motive constructed from M

Chapter 16

$\mathcal{P}_{\mathrm{Ta}}(M)$	Tate periods for M
$\mathcal{P}_2(M)$	periods of the second kind wrt closed paths
$\mathcal{P}_{\mathrm{alg}}(M)$	algebraic periods
$\mathcal{P}_3(M)$	periods of the third kind wrt closed paths
$\mathcal{P}_{\mathrm{inc2}}(M)$	periods of the second kind wrt non-closed paths
$\mathcal{P}_{\mathrm{inc3}}(M)$	periods of the third kind wrt non-closed paths
$\mathcal{P}_{\mathrm{Bk}}(M)$	periods of the third kind wrt non-closed paths in the Baker case
$\delta_?(M)$	dimension of $\mathcal{P}_?(M)$

Chapter 17

Φ	the map $L_{\mathbb{Q}} \otimes X(T)_{\mathbb{Q}} \to \mathcal{P}_{\mathrm{inc3}}(M)$
$R_1(M)$	primitive relations on $L_{\mathbb{Q}} \otimes X(T)_{\mathbb{Q}}$
$R_n(M)$	n-fold relations on $L_{\mathbb{Q}} \otimes X(T)_{\mathbb{Q}}$
$R_{\mathrm{inc2}}(M)$	full relation space

Chapter 18

$\wp(z;\Lambda)$	Weierstraß $\wp$-function for the lattice Λ
g_2, g_3	lattice sums
$\zeta(z;\Lambda)$	Weierstraß ζ-function for the lattice Λ
$\sigma(z;\Lambda)$	Weierstraß σ-function for the lattice λ
ω	standard regular invariant differential on E
η	standard differential of the second kind on E
ξ_P	standard differential with simple pole in P
$F(w,u)$	exponential of ξ_P

Chapter 19

$\xi(\lambda)$	elliptic differential form on $\mathbb{C}$
$\gamma_{p,q}$	path from p to q
$I_{p,q}(\lambda)$	Euler integral from p to q
C_λ	elliptic curve in Legendre normal form
$\omega(\lambda)$	standard invariant differential form on C_λ
$\widetilde{\gamma}_{p,q}$	lift of $\gamma_{p,q}$
$c_{p,q}$	closed path around p and p
$\widetilde{c}_{p,q}$	lift of $c_{p,q}$
$F(1/2,1/2,1;\lambda)$	hypergeometric function
$\mathsf{B}(p,q)$	Euler's Beta-function
C	Legendre family of elliptic curves
$\eta(\lambda)$	standard differential form of the second kind on C_λ
$F(a,b,c;\lambda)$	hypergeometric function in general
$\omega(a,b,c;\lambda)$	differential form in the Euler integral
$\Omega(a,b,c;\lambda)$	Euler integral
$\omega(b,c-b)$	differential form in the Beta integral
$\mathsf{B}(b,c-b)$	Euler's Beta-function
$C_N(\lambda)$	curve with affine equation $y^N = x^r(1-x)^s(1-\lambda x)^t$
C_p	degeneration $C_p(0)$
$X_p(\lambda)$	normalisation of $C_p(\lambda)$
X_p	normalisation of C_p
$J_p(\lambda)$	Jacobian of $X_p(\lambda)$
J_p	Jacobian of X_p
$\omega_n^{u,v,w}$	differential form on $X_p(\lambda)$
$\omega_n^{u,v}$	differential form on X_p
ω_n	$\omega_n^{u,v,w}$ or $\omega_n^{u,v}$ for distinguished choice of u,v,w
$\langle x \rangle$	fractional part of a rational number x

Appendix A

$\mathbb{Q}$-Vect	category of finite-dimensional $\mathbb{Q}$-vector spaces
E-Mod	finitely generated left E-modules
D	diagram, i.e. oriented graph
T	representation of a diagram
Pairs^{eff}	Nori's diagram of (effective) pairs
$\mathcal{MM}_{\text{Nori}}^{\text{eff}}(k, \mathbb{Q})$	category of effective Nori motives
H_{sing}	singular realisation of Nori motives
H_{Nori}	representation of Pairs^{eff} in Nori motives
H_{Hdg}	Hode realisation of Nori motives
H	period realisation of Nori motives
H_{dR}	de Rham realisation of Nori motives
$d_n\mathcal{MM}_{\text{Nori}}(k, \mathbb{Q})$	degree filtration on Nori motives
$\mathbb{Q}(-1)$	Lefschetz motive
$\mathcal{MM}_{\text{Nori}}(k, \mathbb{Q})$	category of all Nori motives
$\mathcal{P}^{\text{eff}}$	set of all periods of effective Nori motives
$E(M)$	endomorphism algebra of $H_{\text{sing}}|_{\langle M \rangle}$

Appendix B

$\text{DM}_{\text{gm}}^{\text{eff}}(k, \mathbb{Q})$	Voevodsky's triangulated category of effective geometric motives
M	standard functor $\text{Var}_k \to \text{DM}_{\text{gm}}^{\text{eff}}(k, \mathbb{Q})$, also representation of Pairs$_k^{\text{eff}}$
H_{sing}	singular realisation of geometric motives
$d_n\text{DM}_{\text{gm}}^{\text{eff}}(k, \mathbb{Q})$	filtration by dimension

Appendix C

V	functor $1\text{-Mot}_k \to (\mathbb{Q}, \overline{\mathbb{Q}})$-Vect
$R\Gamma_{\text{dR}}(X, Y)$	functorial complexes computing relative de Rham cohomology
$R\Gamma_{\text{sing}}(X, Y)$	functorial complexes computing singular cohomology
$R\widetilde{\Gamma}_{\text{dR}}(X, Y, \mathfrak{U})$	Čech complexes computing relative de Rham cohomology
$\underline{G}$	Nisnevich sheaf with transfers defined by G
$\underline{L}$	Nisnevich sheaf with transfers defined by L
$M_{\text{gm}}(\cdot)$	geometric motive defined by a complex of Nisnevich sheaves with transfers
$V(M)^\vee$	external dual of $V(M)$

References

[ABV15] Joseph Ayoub and Luca Barbieri-Viale. Nori 1-motives. *Math. Ann.*, 361(1–2):367–402, 2015.

[ABVB20] F. Andreatta, L. Barbieri-Viale and A. Bertapelle. Motivic periods and Grothendieck arithmetic invariants. *Adv. Math.*, 359:106880 (50pp), 2020. With an appendix by B. Kahn.

[AEWH15] Giuseppe Ancona, Stephen Enright-Ward and Annette Huber. On the motive of a commutative algebraic group. *Doc. Math.*, 20:807–858, 2015.

[Ahl53] Lars V. Ahlfors. *Complex analysis: An introduction to the theory of analytic functions of one complex variable.* McGraw, New York, 1953.

[And96] Yves André. *G*-fonctions et transcendance. *J. Reine Angew. Math.*, 476:95–125, 1996.

[And04] Yves André. *Une introduction aux motifs (motifs purs, motifs mixtes, périodes)*, Panoramas et Synthèses [Panoramas and Syntheses] 17. Société Mathématique de France, Paris, 2004.

[And17] Yves André. Groupes de Galois motiviques et périodes. *Astérisque*, (390):Exp. No. 1104, 1–26, 2017. Séminaire Bourbaki. Vol. 2015/2016. Exposés 1104–1119.

[And19] Yves André. Letter to C. Bertolin, 2019. Appendix to: C. Bertolin, Third kind elliptic integrals and 1-motives. *J. Pure Appl. Alg.*, 224(10):106396, 2020.

[And21] Yves André. A note on 1-motives. *Int. Math. Res. Not. IMRN*, (3):2074–2080, 2021.

[AO18] Masanori Asakura and Noriyuki Otsubo. CM periods, CM regulators and hypergeometric functions, I. *Canad. J. Math.*, 70(3):481–514, 2018.

[Apé79] Roger Apéry. Irrationalité de $\zeta 2$ et $\zeta 3$. *Astérisque*, 61:11–13, 1979. Luminy Conference on Arithmetic.

[Arc03a] Natália Archinard. Exceptional sets of hypergeometric series. *J. Number Theory*, 101(2):244–269, 2003.

[Arc03b] Natália Archinard. Hypergeometric abelian varieties. *Canad. J. Math.*, 55(5):897–932, 2003.

[Arn90] V. I. Arno'd. *Huygens and Barrow, Newton and Hooke: Pioneers in mathematical analysis and catastrophe theory from evolvents to quasicrystals.* Birkhäuser, Basel, 1990. Trans. Eric J. F. Primrose.

[Ayo10] Joseph Ayoub. Note sur les opérations de Grothendieck et la réalisation de Betti. *J. Inst. Math. Jussieu*, 9(2):225–263, 2010.

[Ayo14a] J. Ayoub. Periods and the conjectures of Grothendieck and Kontsevich-Zagier. *Eur. Math. Soc. Newsl.*, 91:12–18, 2014.

[Ayo14b] Joseph Ayoub. La réalisation étale et les opérations de Grothendieck. *Ann. Sci. Éc. Norm. Supér. (4)*, 47(1):1–145, 2014.

[Ayo15] J. Ayoub. Une version relative de la conjecture des périodes de Kontsevich-Zagier. *Ann. of Math. (2)*, 181(3):905–992, 2015.

[Bak66] A. Baker. Linear forms in the logarithms of algebraic numbers I, II, III. *Mathematika*, 13 (1966):204–216; 14 (1967), 102–107; 14 (1967), 220–228; 1966.

[Bak69] A. Baker. On the quasi-periods of the Weierstrass ζ-function. *Nachr. Akad. Wiss. Göttingen Math.-Phys. Kl. II*, 1969:145–157, 1969.

[Bar55] Iacopo Barsotti. Un teorema di struttura per le varietà gruppali. *Atti Accad. Naz. Lincei Rend. Cl. Sci. Fis. Mat. Nat. (8)*, 18:43–50, 1955.

[Ber76] Daniel Bertrand. Séries d'Eisenstein et transcendance. *Bull. Soc. Math. France*, 104(3):309–321, 1976.

[Ber02] Cristiana Bertolin. Périodes de 1-motifs et transcendance. *J. Number Theory*, 97(2):204–221, 2002.

[Ber20] Cristiana Bertolin. Third kind elliptic integrals and 1-motives. *J. Pure Appl. Algebra*, 224(10):106396, 28pp, 2020. With a letter of Y. André and an appendix by M. Waldschmidt.

[Bos21] Alin Bostan. A proof of Archinard's identity, 2020/21. Email to the authors.

[BR01] Keith Ball and Tanguy Rivoal. Irrationalité d'une infinité de valeurs de la fonction zêta aux entiers impairs. *Invent. Math.*, 146(1):193–207, 2001.

[Bri17] Michel Brion. Commutative algebraic groups up to isogeny. *Doc. Math.*, 22:679–725, 2017.

[Bro12] F. Brown. Mixed Tate motives over $\mathbb{Z}$. *Ann. of Math. (2)*, 175(2):949–976, 2012.

[BVK16] Luca Barbieri-Viale and Bruno Kahn. On the derived category of 1-motives. *Astérisque*, 381: 254pp, 2016.

[BW07] A. Baker and G. Wüstholz. *Logarithmic forms and Diophantine geometry*, New Mathematical Monographs 9. Cambridge University Press, Cambridge, 2007.

[Car35] Leonard Carlitz. On certain functions connected with polynomials in a Galois field. *Duke Math. J.*, 1(2):137–168, 1935.

[CC88] D. V. Chudnovsky and G. V. Chudnovsky. Approximations and complex multiplication according to Ramanujan. In *Ramanujan revisited (Urbana-Champaign, Ill., 1987)*, pp. 375–472. Academic Press, Boston, MA, 1988.

[Cha85] K. Chandrasekharan. *Elliptic functions*, Grundlehren der Mathematischen Wissenschaften [Fundamental Principles of Mathematical Sciences] 281. Springer-Verlag, Berlin, 1985.

[Cha17] Chieh-Yu Chang. Periods, logarithms and multiple zeta values, 2017. To appear: Proceedings of first annual meeting of ICCM.

[Che60] C. Chevalley. Une démonstration d'un théorème sur les groupes algébriques. *J. Math. Pures Appl. (9)*, 39:307–317, 1960.

[Chu80]   G. V. Chudnovsky. Algebraic independence of values of exponential and elliptic functions. In *Proceedings of the International Congress of Mathematicians (Helsinki, 1978)*, pp. 339–350. Acad. Sci. Fennica, Helsinki, 1980.

[CPTY10] Chieh-Yu Chang, Matthew A. Papanikolas, Dinesh S. Thakur and Jing Yu. Algebraic independence of arithmetic gamma values and Carlitz zeta values. *Adv. Math.*, 223(4):1137–1154, 2010.

[CY07]   Chieh-Yu Chang and Jing Yu. Determination of algebraic relations among special zeta values in positive characteristic. *Adv. Math.*, 216(1):321–345, 2007.

[Del71]   Pierre Deligne. Théorie de Hodge. II. *Inst. Hautes Études Sci. Publ. Math.*, 40:5–57, 1971.

[Del72]   Pierre Deligne. Les intersections complètes de niveau de Hodge un. *Invent. Math.*, 15:237–250, 1972.

[Del74]   Pierre Deligne. Théorie de Hodge. III. *Inst. Hautes Études Sci. Publ. Math.*, 44:5–77, 1974.

[DG70]   Michel Demazure and Pierre Gabriel. *Groupes algébriques. Tome I: Géométrie algébrique, généralités, groupes commutatifs*. Masson & Cie, Éditeur, Paris; North-Holland Publishing Co., Amsterdam, 1970. Avec un appendice, *Corps de classes local* par Michiel Hazewinkel.

[DG05]   P. Deligne and A. B. Goncharov. Groupes fondamentaux motiviques de Tate mixte. *Ann. Sci. École Norm. Sup. (4)*, 38(1):1–56, 2005.

[DM82]   P. Deligne and J. Milne. Tannakian categories. In Pierre Deligne, James S. Milne, Arthur Ogus and Kuang-yen Shih (eds), *Hodge cycles, motives, and Shimura varieties*, Lecture Notes in Mathematics 900, pp. 101–228. Springer-Verlag, Berlin–New York, 1982.

[Fri11]   Robert Fricke. *Die elliptischen Funktionen und ihre Anwendungen. Erster Teil. Die funktionentheoretischen und analytischen Grundlagen*. Springer, Heidelberg, 2011. Reprint of the 1916 original, With a foreword by the editors of Part III: Clemens Adelmann, Jürgen Elstrodt and Elena Klimenko.

[Gel34]   A.O. Gelfond. Sur le septième problème de Hilbert. *Dokl. Akad. Nauk 2, Izvest. Akad. Nauk SSSR*, pp. 623–630, 1934.

[GH78]   P. Griffiths and J. Harris. *Principles of algebraic geometry*. Pure and Applied Mathematics. Wiley-Interscience, New York, 1978.

[GR78]   Benedict H. Gross and David E. Rohrlich. Some results on the Mordell–Weil group of the Jacobian of the Fermat curve. *Invent. Math.*, 44(3):201–224, 1978.

[Gro66]   Alexandre Grothendieck. On the de Rham cohomology of algebraic varieties. *Inst. Hautes Études Sci. Publ. Math.*, 29:95–103, 1966.

[Gro79]   B. H. Gross. On an identity of Chowla and Selberg. *J. Number Theory*, 11(3):344–348, 1979.

[Gro20]   Benedict H. Gross. On the periods of Abelian varieties. *ICCM Not.*, 8(2):10–18, 2020.

[Har75]   R. Hartshorne. On the de Rham cohomology of algebraic varieties. *Inst. Hautes Études Sci. Publ. Math.*, 45:5–99, 1975.

[Har16] Daniel Harrer. Comparison of the categories of motives defined by Voevodsky and Nori. PhD thesis, Albert-Ludwigs-Universität Freiburg, 2016. arXiv:1609.05516.

[Hat02] A. Hatcher. *Algebraic topology*. Cambridge University Press, Cambridge, 2002.

[HMS17] Annette Huber and Stefan Müller-Stach. *Periods and Nori motives*, Ergebnisse der Mathematik und ihrer Grenzgebiete. 3. Folge. A Series of Modern Surveys in Mathematics [Results in Mathematics and Related Areas. 3rd Series. A Series of Modern Surveys in Mathematics] 65. Springer, Cham, 2017. With contributions by Benjamin Friedrich and Jonas von Wangenheim.

[Hör21] Fritz Hörmann. A note on formal periods, 2021. arXiv:2106.03803.

[Hub95] Annette Huber. *Mixed motives and their realization in derived categories*, Lecture Notes in Mathematics 1604. Springer-Verlag, Berlin, 1995.

[Hub00] Annette Huber. Realization of Voevodsky's motives. *J. Algebraic Geom.*, 9(4):755–799, 2000.

[Hub04] Annette Huber. Corrigendum to: 'Realization of Voevodsky's motives' [*J. Algebraic Geom.* 9(4):755–799, 2000; MR1775312]. *J. Algebraic Geom.*, 13(1):195–207, 2004.

[Hub20] Annette Huber. Galois theory of periods. *Münster J. Math.*, 13(2):573–596, 2020.

[IKSY91] Katsunori Iwasaki, Hironobu Kimura, Shun Shimomura, and Masaaki Yoshida. *From Gauss to Painlevé: A modern theory of special functions*, Aspects of Mathematics 16. Friedr. Vieweg & Sohn, Braunschweig, 1991.

[Kle81] Felix Klein. *Vorlesungen über die hypergeometrische Funktion*, Grundlehren der Mathematischen Wissenschaften [Fundamental Principles of Mathematical Sciences] 39. Springer-Verlag, Berlin–New York, 1981. Reprint of the 1933 original.

[Kon99] Maxim Kontsevich. Operads and motives in deformation quantization. *Lett. Math. Phys.*, 48(1):35–72, 1999.

[KS90] M. Kashiwara and P. Schapira. *Sheaves on manifolds*, Grundlehren der Mathematischen Wissenschaften [Fundamental Principles of Mathematical Sciences] 292. Springer-Verlag, Berlin, 1990. With a chapter in French by Christian Houzel.

[KZ01] Maxim Kontsevich and Don Zagier. Periods. In Björn Engquist and Wilfried Schmid (eds), *Mathematics unlimited – 2001 and beyond*, pp. 771–808. Springer, Berlin, 2001.

[Mil08] J. Milne. Abelian varieties, 2008. Available at www.jmilne.org/math/CourseNotes/AV.pdf.

[Mum70] David Mumford. *Abelian varieties*. Tata Institute of Fundamental Research Studies in Mathematics 5. Published for the Tata Institute of Fundamental Research, Bombay. Oxford University Press, London, 1970.

[MW93] David Masser and Gisbert Wüstholz. Isogeny estimates for abelian varieties, and finiteness theorems. *Ann. of Math. (2)*, 137(3):459–472, 1993.

[Org04] Fabrice Orgogozo. Isomotifs de dimension inférieure ou égale à un. *Manuscripta Math.*, 115(3):339–360, 2004.

[Sch34a] Th. Schneider. Transzendenzuntersuchungen periodischer Funktionen II. Transzendenzeigenschaften elliptischer Funktionen. *J. Reine Angew. Math.*, 172:70–74, 1934.

[Sch34b] Theodor Schneider. Transzendenzuntersuchungen periodischer Funktionen I. Transzendenz von Potenzen. *J. Reine Angew. Math.*, 172:65–69, 1934.

[Sch37] Theodor Schneider. Arithmetische Untersuchungen elliptischer Integrale. *Math. Ann.*, 113(1):1–13, 1937.

[Sch57] Theodor Schneider. *Einführung in die transzendenten Zahlen*, Vol. 81. Springer, Berlin, 1957.

[Ser60] Jean-Pierre Serre. Morphismes universels et différentielles de troisième espèce. *Séminaire Claude Chevalley* 4 (1958–1959): 1–8, 1960. Available at <http://eudml.org/doc/110340>.

[Ser88] Jean-Pierre Serre. *Algebraic groups and class fields*, Graduate Texts in Mathematics 117. Springer-Verlag, New York, 1988. Translated from the French.

[Sie32] C. L. Siegel. Über die Perioden elliptischer Funktionen. *J. Reine Angew. Math.*, 167:62–69, 1932.

[Sie49] Carl Ludwig Siegel. *Transcendental numbers*. Annals of Mathematics Studies 16. Princeton University Press, Princeton, NJ, 1949.

[Spa66] E. H. Spanier. *Algebraic topology*. McGraw-Hill, New York, 1966.

[Tre17] Paula Tretkoff. *Periods and special functions in transcendence*. Advanced Textbooks in Mathematics. World Scientific, Hackensack, NJ, 2017.

[Tsi18] Jacob Tsimerman. The André–Oort conjecture for $\mathcal{A}_g$. *Ann. of Math. (2)*, 187(2):379–390, 2018.

[VdP71] A. J. Van der Poorten. On the arithmetic nature of definite integrals of rational functions. *Proc. Amer. Math. Soc.*, 29:451–456, 1971.

[Voe00] Vladimir Voevodsky. Triangulated categories of motives over a field. In Vladimir Voevodsky, Andrei Suslin and Eric M. Firedlander (eds), *Cycles, transfers, and motivic homology theories*, Annals of Mathematics Studies 143, pp. 188–238. Princeton University Press, Princeton, NJ, 2000.

[Wad46] L. I. Wade. Transcendence properties of the Carlitz ψ-functions. *Duke Math. J.*, 13:79–85, 1946.

[War83] F. W. Warner. *Foundations of differentiable manifolds and Lie groups*, Graduate Texts in Mathematics 94. Springer-Verlag, New York-Berlin, 1983. Corrected reprint of the 1971 edition.

[Wei85] C. Weierstrass. Zu Lindemann's Abhandlung: 'Über die Ludolph'sche Zahl'. *Berl. Ber.*, 1885:1067–1086, 1885.

[Wel17] Raymond O. Wells, Jr. *Differential and complex geometry: Origins, abstractions and embeddings*. Springer, Cham, 2017.

[Wol88] Jürgen Wolfart. Werte hypergeometrischer Funktionen. *Invent. Math.*, 92(1):187–216, 1988.

[Wüs84a] G. Wüstholz. Recent progress in transcendence theory. In *Number theory (Noordwijkerhout, 1983)*, Lecture Notes in Mathematics 1068, pp. 280–296. Springer, Berlin, 1984.

[Wüs84b] Gisbert Wüstholz. Transzendenzeigenschaften von Perioden elliptischer Integrale. *J. Reine Angew. Math.*, 354:164–174, 1984.

[Wüs87] Gisbert Wüstholz. Algebraic groups, Hodge theory, and transcendence. In *Proceedings of the International Congress of Mathematicians, Vols 1, 2 (Berkeley, Calif., 1986)*, pp. 476–483. American Mathematical Society, Providence, RI, 1987.

[Wüs89] Gisbert Wüstholz. Algebraische Punkte auf analytischen Untergruppen algebraischer Gruppen. *Ann. of Math. (2)*, 129(3):501–517, 1989.

[Wüs12] Gisbert Wüstholz. Leibniz' conjecture, periods and motives. In *Colloquium De Giorgi 2009*, Colloquia 3, pp. 33–42. Edizioni della Normale, Pisa, 2012.

[Wüs21] Gisbert Wüstholz. Elliptic and abelian period spaces. *Acta Arithmetica*, 189:329–357, 2021.

[WW85] Jürgen Wolfart and Gisbert Wüstholz. Der Überlagerungsradius gewisser algebraischer Kurven und die Werte der Betafunktion an rationalen Stellen. *Math. Ann.*, 273(1):1–15, 1985.

[Yu97] Jing Yu. Analytic homomorphisms into Drinfeld modules. *Ann. of Math. (2)*, 145(2):215–233, 1997.

[Zud01] V. V. Zudilin. One of the numbers $\zeta(5)$, $\zeta(7)$, $\zeta(9)$, $\zeta(11)$ is irrational. *Uspekhi Mat. Nauk*, 56(4(340)):149–150, 2001.

Index

241

Printed in the United States
by Baker & Taylor Publisher Services